T0269177

NEURONAL CORRELATES OF EMPATHY

NEURONAL CORRELATES OF EMPATHY

FROM RODENT TO HUMAN

Edited by

KSENIA Z. MEYZA

Assistant Professor and Associate Researcher, Laboratory of Emotions Neurobiology,
Nencki Institute of Experimental Biology at the Polish Academy of Sciences, Warsaw, Poland

EWELINA KNAPSKA

Associate Professor and Head, Laboratory of Emotions Neurobiology,
Nencki Institute of Experimental Biology at the Polish Academy of Sciences, Warsaw, Poland

ACADEMIC PRESS

An imprint of Elsevier

Academic Press is an imprint of Elsevier
125 London Wall, London EC2Y 5AS, United Kingdom
525 B Street, Suite 1800, San Diego, CA 92101-4495, United States
50 Hampshire Street, 5th Floor, Cambridge, MA 02139, United States
The Boulevard, Langford Lane, Kidlington, Oxford OX5 1GB, United Kingdom

Library of Congress Cataloging-in-Publication Data
A catalog record for this book is available from the Library of Congress

British Library Cataloguing-in-Publication Data
A catalogue record for this book is available from the British Library

ISBN: 978-0-12-805397-3

For information on all Academic Press publications visit our website at
https://www.elsevier.com/books-and-journals

  Working together
to grow libraries in
developing countries

www.elsevier.com • www.bookaid.org

Publisher: Nikki Levy
Acquisition Editor: Joslyn Chaiprasert-Paguio
Editorial Project Manager: Timothy Bennett
Production Project Manager: Mohana Natarajan
Designer: Christian Bilbow

Typeset by Thomson Digital

Dedication

Jaak Panksepp, who died at the age of 73 on April 18, 2017, was an Estonian-American neurobiologist of extraordinary importance for affective neuroscience, a field that he founded. He placed human and animal emotions on a continuum, and was the first to develop a neuroscience of the emotions. He had to fight many establishment forces, the most resistant one being radical behaviorism, which considered human emotions irrelevant, and animal emotions suspect. He did more than almost any scientist to make animal emotions a respectable topic of discourse, and became especially known for his studies of joy, play, and laughter in rats by tickling his subjects and recording their ultrasonic vocalizations. But his work went far beyond this, and situated the emotions in ancient subcortical brain areas rather than the cerebral cortex. He distinguished seven primal emotions, considering them homologous across mammals, and far less subject to mental construction than some psychologists claim. I have seen Panksepp defend this position, bringing a passion and knowledge to it that few could match. He did not limit himself to observable behavior, as the behaviorists would like, but assumed that emotions always come with subjective experiences, so that they are felt, even by rats. His magnum opus, *Affective Neuroscience* (1998), became a bestseller by academic standards, and will remain a standby of the field for years to come. Panksepp was ahead of his time, and influenced many. He will be greatly missed.

Contents

9. Neuronal Correlates of Remote Fear Learning in Rats

KAROLINA ROKOSZ, EWELINA KNAPSKA

10. Lost in Translation: Improving Our Understanding of Pain Empathy

SIVAANI SIVASELVACHANDRAN, MERUBA SIVASELVACHANDRAN,
SALSABIL ABDALLAH, LOREN J. MARTIN

11. Relief From Stress Provided by Conspecifics: Social Buffering

YASUSHI KIYOKAWA

12. Helping Behavior in Rats

INBAL BEN-AMI BARTAL, PEGGY MASON

13. Challenging Convention in Empathy Research: Developing a Mouse Model and Initial Neural Analyses

JULES B. PANKSEPP, GARET P. LAHVIS

14. Lack of Empathy—Mouse Models

KSENIA Z. MEYZA

15. Future Directions, Outstanding Questions

EWELINA KNAPSKA, KSENIA Z. MEYZA

List of Contributors

Salsabil Abdallah   University of Toronto Mississauga, Mississauga, ON, Canada

Inbal Ben-Ami Bartal   University of Chicago, Chicago, IL, United States

Zanna Clay   University of Birmingham, Birmingham, United Kingdom

Gino Coudé   Institut des Sciences Cognitives Marc Jeannerod, CNRS, Cedex, France

Frans B.M. de Waal   Emory University, Atlanta, GA, United States; Living Links, Yerkes National Primate Research Center, Atlanta, GA, United States

Pier F. Ferrari   Institut des Sciences Cognitives Marc Jeannerod, CNRS, Cedex, France

Valeria Gazzola   Netherlands Institute for Neuroscience, an Institute of the Royal Netherlands Academy of Art and Sciences (KNAW), Amsterdam, The Netherlands; University of Amsterdam, Amsterdam, The Netherlands

Carolyn E. Jones   The University of Texas at Austin, Austin, TX, United States

Christian Keysers   Netherlands Institute for Neuroscience, an Institute of the Royal Netherlands Academy of Art and Sciences (KNAW), Amsterdam, The Netherlands; University of Amsterdam, Amsterdam, The Netherlands

Yasushi Kiyokawa   The University of Tokyo, Tokyo, Japan

Ewelina Knapska   Neurobiology of Emotions Laboratory, Nencki Institute of Experimental Biology of the Polish Academy of Sciences, Warsaw, Poland

Garet P. Lahvis   Oregon Health and Science University, Portland, OR, United States

Claus Lamm   University of Vienna, Vienna, Austria

Loren J. Martin   University of Toronto Mississauga, Mississauga, ON, Canada

Peggy Mason   University of Chicago, Chicago, IL, United States

Ksenia Z. Meyza   Nencki Institute of Experimental Biology of the Polish Academy of Sciences, Warsaw, Poland

Marie-H. Monfils   The University of Texas at Austin, Austin, TX, United States

Andreas Olsson   Karolinska Institutet, Solna, Sweden

Elisabetta Palagi   Natural History Museum, University of Pisa, Pisa, Italy; Institute of Cognitive Sciences and Technologies (CNR Rome), Italy

Jules B. Panksepp   Oregon Health and Science University, Portland, OR, United States

Karolina Rokosz   Neurobiology of Emotions Laboratory, Nencki Institute of Experimental Biology of the Polish Academy of Sciences, Warsaw, Poland

Meruba Sivaselvachandran   University of Toronto Mississauga, Mississauga, ON, Canada

Sivaani Sivaselvachandran   University of Toronto Mississauga, Mississauga, ON, Canada

Adam S. Smith   Pharmacy School, University of Kansas, Lawrence, KS, United States

Victoria Spring   Pennsylvania State University, State College, PA, United States

Livia Tomova   University of Vienna, Vienna, Austria

Zuoxin Wang   Florida State University, Tallahassee, FL, United States

Preface

In recent years, we have become accustomed to hearing questions, such as "Empathy in animals? Seriously?" when discussing with people what we do in the laboratory. Although initially frustrating to have to repeatedly explain that empathy is an evolutionarily continuous phenomenon, it has also made us realize that there is a big gap between the common understanding of what empathy is and actual knowledge of the mechanisms involved. Part of the blame lies with scientists, as we seem not to have been able to come up with a uniform definition of empathy or related phenomena. There is an ongoing dispute about which processes form empathy (by some labeled "empathy proper") and which phenomena are only loosely associated with it. The divide is further deepened by the level of anthropomorphism used in defining empathy, which differs greatly between neuroscientists and psychologists. Much effort is needed on all sides to bridge that gap.

The idea for the book was born after the symposium "Neuronal correlates of rodent empathy," held at the 24th Annual Meeting of the International Behavioral Neuroscience Society. Already at the commencement of the project, we talked about expanding the material describing neuronal correlates of empathic behaviors to species other than mice and rats together with the speakers of the symposium Dr. Inbal Bartal Ben-Ami, Dr. Marie Monfils, and Dr. Jules Panksepp. By including other species and phenomena, such as social buffering and pair bonding, as well as a discussion of factors influencing our innate empathic traits, we created what we believe is a source of comprehensive information for anyone interested in biological mechanisms of empathic behaviors. We hope that by bringing psychologists, behaviorists, and neuroscientists together, this book will facilitate exchange of knowledge among various specialists and thereby serve as a first step to close the gap between their differing views of empathy.

1

Introduction–Empathy Beyond Semantics

Ewelina Knapska, Ksenia Z. Meyza

Nencki Institute of Experimental Biology, Polish Academy of Sciences, Warsaw, Poland

From the second half of the 18th and throughout the 19th century, philosophers discussed our ability to "feel into" works of art and nature as an explanatory account of the phenomenological immediacy of our aesthetic experiences. According to these theories, aesthetic appreciation of objects was achieved by projecting one's own imagined feelings onto the world (https://plato.stanford.edu/entries/empathy/). It is, however, a German philosopher, Theodor Lipps (1851–1914), who is remembered as the father of the first scientific theory of *Einfühlung* (literally meaning "feeling-into"). Lipps broadened the meaning of this term from a concept of philosophical aesthetics to a central category of social sciences by explaining how people understand the mental states of others. He adapted Hume's concept of "sympathy," a process that allows the contents of "the minds of men" to become "mirrors to one another". According to Lipps, the unconscious process of *Einfühlung*, entailed a sense of merging the observer with the observed (Montag, Gallinat, & Heinz, 2008). The resonance was achieved by "inner imitation" based on an innate disposition for motor mimicry, that is, triggering processes that give rise to similar kinaesthetic sensations in both the observer and the observed target. Lipps regarded recognition, not only of emotions expressed in bodily gestures or facial expressions, but of all mental activities (intellectual empathy) as being based on inner imitation. Although considered speculative in his time, Lipps's theory of "inner imitation" has some reflection in present-day concepts to be discussed in detail later.

The term *Einfühlung*, translated as *empathy* from the Greek ἐμπάθεια (*empatheia*) and literally meaning ἐν (*en*), "in/at" + πάθος (*pathos*) "passion/suffering," was introduced into the English language by the psychologist

Neuronal Correlates of Empathy. http://dx.doi.org/10.1016/B978-0-12-805397-3.00001-2

Edward Titchener in 1909. As its debut in the English language, empathy has been discussed mainly from the clinical perspective of nursing, psychotherapy, and psychiatry and has received more attention from health care professionals and philosophers than neuropsychologists. Social cognition in humans, the psychological processes that allow us to make inferences about other people's intentions, feelings, and thoughts, did not begin to attract the attention of researchers until the 1980s. A revival of scientific interest in social psychological phenomena started in 1985 with the publication of "Social Brain. Discovering the Networks of the Mind." by Michael Gazzaniga (Gazzaniga, 1985). This was the first modern attempt to link social behavior with the function of the brain.

In psychology, empathy has been traditionally a subject of study in the domain of *social cognitive neuroscience* rather than social neuroscience. The latter is centered on understanding the brain structures involved in social motivation including the amygdala, hypothalamus, brainstem, and basal ganglia, rather than on cognitive processing. Social cognitive neuroscience, on the other hand, deals with higher-order cognitive processes found predominantly in humans and nonhuman primates and is related to associative cortical areas (Easton & Emery, 2004). The exploration of neuronal mechanisms of social interactions started with single-unit recordings in the cortex of primates. Studies in the 1980s and 90s produced several important discoveries: for instance, identification of neurons in the anterior temporal cortex that were selective to social stimuli, such as faces (Bruce, Desimone, & Gross, 1981; Perrett, Rolls, & Caan, 1982) and neurons in the superior temporal sulcus that responded to the presence of socially significant motion, such as eye gaze movement (Perrett et al., 1989). Giacomo Rizzolatti's group, conducting single-unit recordings from macaque premotor and inferior parietal cortices, discovered *mirror neurons*, which responded both when the macaque performed an action and during observation of the same action being performed by the experimenter (Gallese, Fadiga, Fogassi, & Rizzolatti, 1996; Rizzolatti, Fadiga, Gallese, & Fogassi, 1996; Umiltà et al., 2001). The discovery of mirror neurons fueled speculation about the neuronal mechanisms of imitation and mimicry. The involvement of a mirroring mechanism was hypothesized in a wide range of abilities, including empathy (Baird, Scheffer, & Wilson, 2011). According to this hypothesis, we gain insight into the feelings of other people by virtue of vicarious activations (also referred to as "shared representations"), where the same neurons activated by directly experienced emotions are also triggered by observing/interacting with others who are experiencing or communicating a recent experience of these emotions. In the early 2000s, vigorous development of human brain imaging techniques [especially functional magnetic resonance imaging (fMRI)] encouraged systematic neuropsychological studies of empathy that provided several examples of vicarious activation in a subject when they were observing the emotions of others (Bernhardt & Singer, 2012; Stanley & Adolphs, 2013). For instance, both people feeling disgust and observing faces expressing disgust display activation in the anterior insula and, to a lesser extent, the anterior cingulate cortex (Wicker et al., 2003). Similarly, Singer and coworkers (Singer et al., 2004) showed that experiencing pain and empathizing with the pain of others evoked overlapping neural activations in the cingulate and insular cortices.

The impact of human neuroimaging studies on our understanding of neurobiological correlates of empathy is tremendous. Apart from identifying empathy relevant brain regions, they have allowed observation of the dynamic interactions of distinct parts of brain circuitry in real time. Such studies also illuminate the natural spectrum of individual differences in empathic responsiveness, including specific subpopulations of subjects with preexisting conditions affecting that reactivity. They also allow for longitudinal studies.

Although we know much more about the neuronal underpinnings of empathy than we did 30 years ago, there are still crucial, unanswered questions ahead of us. One of the most important among them, despite many years of studies, concerns the function of mirror neurons and their role in vicarious activations (Hickok, 2009). The main issue that needs to be addressed is the limited resolution and correlative nature of human neuroimaging studies. Gaining mechanistic insight into the exquisite organization of the neuronal circuitry underlying empathic behaviors requires manipulation of the neuronal activity. Successful protocols for such manipulation (described later) are now routinely used in rodents. Their use in primates, however, is severely limited by technical and ethical constraints. Although animal studies shed light on the neural basis of human social behavior, their use is often considered problematic. This is mainly due to human social behavior being extremely complex as compared with that of other species used in empathy research. Later, we discuss these problems and their potential solutions.

Contemporary researchers define human empathy as the capacity to understand the behavior of other people by inferring their mental states and to respond with an appropriate emotion. Investigation of the neuronal underpinnings of such a complex phenomenon would be impossible without breaking it down into biologically based subprocesses. One such division was proposed by Preston and de Waal (2002). According to their theory, empathy should be considered a multilayered entity extending from simple forms of emotional contagion to complex forms of cognitive perspective taking. This *multilevel conceptualization of empathy* puts the simplest forms of empathy, involving adoption of another's emotional state (*emotional contagion*) at the core of all empathic behaviors. This relatively simple phenomenon is then followed by more complex levels of the continuum, which might include concern about the state of another individual, attempts to ameliorate this state through consolation (*sympathetic concern*), and/or attribution of the emotional state to someone other than oneself (*empathetic perspective taking*).

The last years have seen the birth of several inclusive theories of empathy, which place the phenomenon in a family of related processes, such as emotional contagion, empathy proper, and sympathy (Chapter 4). They moved studies of empathy from the domain of social cognitive neuroscience to the broader perspective of social neuroscience. Accumulating data, showing emotional contagion and prosocial behaviors in multiple species, including rodents (Chapters 5 through 14), suggests that some widely defined forms of empathy are indeed phylogenetically older than humans. Acknowledging the evolutionary roots of empathy and the existence of some of its forms in animals other than humans forms a framework for studying the primal emotional foundations of empathy in the mammalian brains (Anderson & Adolphs, 2014). As neural mechanisms of empathy are largely unknown, and, to date, no methods allowing detailed insight into the mechanism of such control are available for human studies, animal models open a very interesting and potentially fruitful path for research. However, full agreement on the definition of empathy has not yet been reached. Some psychologists define empathy as a phenomenon requiring conscious awareness of the source of the evoked emotions, thus making it a uniquely human trait. They also distinguish emotional contagion, sympathy, compassion, and prosocial behavior from empathy, arguing that these are neither necessary nor sufficient for the experience of empathy (Chapters 3).

The problems with defining empathy will probably not be solved until we learn the brain mechanisms underlying emotional contagion, sympathetic concern, and empathetic perspective taking, and define the mechanistic relationships between these phenomena. The newly developed technologies for manipulation of neuronal activity, now routinely used in rodent studies, bring about unique opportunities to investigate empathy-relevant neural circuitry with unprecedented detail. Genetic and viral tools, optogenetics, and advanced in vivo imaging techniques currently allow for identification of distinct neuronal circuits that, as the research of the last decade has shown, can control different, often opposite, behaviors while being physically located in the same brain structure (Tovote, Fadok, & Lüthi, 2015). This discovery moved the focus of interest in neurobiology from structurally defined brain areas to neuronal circuits within these structures. In the face of this shift of interest, studying the neuronal mechanisms of empathic behaviors using animal models seems especially important. In fact, to fully understand the neural processes underlying empathy, comparative studies need to be carried out on both human and nonhuman animal species. Obviously, the translational value of such animal research or the degree of functional similarity between the neurocognitive processes observed in humans and other animal species, must be carefully assessed (Chapter 10). Such a comparative approach to empathy may, however, answer many questions about the nature and the evolutionary history of the phenomenon.

One of the aims of this book is to emphasize and reflect on the evolutionary continuity of empathic abilities. To do so, we will navigate between human studies, primate research, and studies on rodents. The latter include studies on prairie voles, as well as rats and mice. The order in which the chapters are presented reflects the gradual unfolding of the evolutionary roots of empathy, thus giving the reader the opportunity to relate first to their own empathic abilities and see them through the lens of proposed definitions of empathy and related phenomena (Chapters 2, 3 and 4). Later, the reader is provided with examples of empathic behaviors in primates (Chapter 5), followed by a detailed description of the Mirror Neuron System found in primates (Chapter 6). Next, the chapter on prairie voles elaborates on the neuronal background of pair bonding and prosocial behaviors in this monogamous species (Chapter 7). The following chapters explore distinct, empathy-related behavioral protocols used in rat (Chapters 8, 9, 11 and 12) and mouse (Chapters 10, 13 and 14) empathy research. These include: Fear by Proxy (Chapter 8), Socially Transferred Fear (Chapter 9), Observational Fear (Chapters 7 and 13), and Shared Pain Experience (Chapter 10) paradigms. Prosocial behavior of rats will be described in detail in Chapters 11 and 12. Further insights into the bottom-up neuronal control of empathic responses are given in Chapter 13, while translational aspects of rodent empathy research are covered in Chapter 14.

It is also our aim to present the state of art knowledge in the many areas of empathy research reviewed in this book, to a broad audience with the hope of drawing further interest to this new, dynamically developing research field. This interest and enthusiasm is much needed, as the search for the ultimate neuronal correlates of empathy is still very much in its infancy. We also hope that this book will establish a base for a dialogue between specialists in human empathy and non-human animal researchers, by providing common definitions of empathy-related phenomena (https://plato.stanford.edu/entries/empathy/).

References

Anderson, D. J., & Adolphs, R. (2014). A framework for studying emotions across species. *Cell, 157*(1), 187–200.

Baird, A. D., Scheffer, I. E., & Wilson, S. J. (2011). Mirror neuron system involvement in empathy: a critical look at the evidence. *Social Neuroscience, 6*(4), 327–335.

Bernhardt, B. C., & Singer, T. (2012). The neural basis of empathy. *Annual Review of Neuroscience, 35*, 1–23.

Bruce, C., Desimone, R., & Gross, C. G. (1981). Visual properties of neurons in a polysensory area in superior temporal sulcus of the macaque. *Journal of Neurophysiology, 46*(2), 369–384.

Easton, A., & Emery, N. (2004). *The cognitive neuroscience of social behaviour.* Hove/New York: Psychology Press.

Gallese, V., Fadiga, L., Fogassi, L., & Rizzolatti, G. (1996). Action recognition in the premotor cortex. *Brain, 119*(Pt. 2), 593–609.

Gazzaniga, M. (1985). *Social brain. Discovering the networks of the Mind.* New York: Basic Books.

Hickok, G. (2009). Eight problems for the mirror neuron theory of action understanding in monkeys and humans. *Journal of Cognitive Neuroscience, 21*(7), 1229–1243.

Montag, C., Gallinat, J., & Heinz, A. (2008). Theodor Lipps and the concept of empathy: 1851–1914. *American Journal of Psychiatry, 165*(10), 1261.

Perrett, D. I., Harries, M. H., Bevan, R., Thomas, S., Benson, P. J., Mistlin, A. J., Chitty, A. J., Hietanen, J. K., & Ortega, J. E. (1989). Frameworks of analysis for the neural representation of animate objects and actions. *Journal of Experimental Biology, 146*, 87–113.

Perrett, D. I., Rolls, E. T., & Caan, W. (1982). Visual neurones responsive to faces in the monkey temporal cortex. *Experimental Brain Research, 47*(3), 329–342.

Preston, S. D., & de Waal, F. B. (2002). Empathy: its ultimate and proximate bases. *Behavioral and Brain Sciences, 25*(1), 1–20.

Rizzolatti, G., Fadiga, L., Gallese, V., & Fogassi, L. (1996). Premotor cortex and the recognition of motor actions. *Brain Research Cognitive Brain Research, 3*(2), 131–141.

Singer, T., Seymour, B., O'Doherty, J., Kaube, H., Dolan, R. J., & Frith, C. D. (2004). Empathy for pain involves the affective but not sensory components of pain. *Science, 303*(5661), 1157–1162.

Stanley, D. A., & Adolphs, R. (2013). Toward a neural basis for social behavior. *Neuron, 80*(3), 816–826.

Tovote, P., Fadok, J. P., & Lüthi, A. (2015). Neuronal circuits for fear and anxiety. *Nature Reviews Neuroscience, 16*(6), 317–331.

Umiltà, M. A., Kohler, E., Gallese, V., Fogassi, L., Fadiga, L., Keysers, C., & Rizzolatti, G. (2001). I know what you are doing. a neurophysiological study. *Neuron, 31*(1), 155–165.

Wicker, B., Keysers, C., Plailly, J., Royet, J. P., Gallese, V., & Rizzolatti, G. (2003). Both of us disgusted in My insula: the common neural basis of seeing and feeling disgust. *Neuron, 40*(3), 655–664.

Further Readings

de Waal, F. B. (2008). Putting the altruism back into altruism: the evolution of empathy. *Annual Review of Psychology, 59*, 279–300.

Singer, T. (2006). The neuronal basis and ontogeny of empathy and mind reading: review of literature and implications for future research. *Neuroscience & Biobehavioral Reviews, 30*(6), 855–863.

2

The Vicarious Brain: Integrating Empathy and Emotional Learning

Andreas Olsson*, Victoria Spring**

*Karolinska Institutet, Stockholm, Sweden; **Pennsylvania State University, State College, PA, United States

INTRODUCTION

Information about what should be approached and avoided is the core currency of survival. Such information can be gained by directly interacting with the world, for example, through trial and error. However, in social species, such as ours, information can also be gleaned indirectly, for example, by observing and communicating with others. Such socially transmitted or "vicarious" emotional learning is often both faster and safer than learning through direct experiences (Bandura, 1979; Laland, 2004; Olsson & Phelps, 2007). For instance, hearing about or witnessing someone else's fear and pain after being harassed in a specific neighborhood motivates learned avoidance of the area, saving you from future harmful personal encounters. As time passes, new vicarious and/or direct experiences might update your impression of the neighborhood, and your likelihood ratio of being personally attacked. New learning takes place.

Apart from allowing you to learn about opportunities and challenges in the environment, others' emotional expressions can also initiate empathic responses; here, referring to the understanding and sharing others' emotional experiences. Empathic responses also motivate actions, for example, alleviating the others' pain or rejoicing in their happiness. In this way, empathizing with others establishes and maintains important social bonds between children and caretakers, friends, coalitional partners, and,

Neuronal Correlates of Empathy. http://dx.doi.org/10.1016/B978-0-12-805397-3.00002-4

depending on the circumstances, between an expanding circle of unrelated individuals (Morelli, Lieberman, & Zaki, 2015; Singer, 2011).

In sum, information about others' emotional experiences can initiate both learning and empathic processes. But how are these processes related to each other? In this chapter, we will first discuss how learning from others' emotional expressions can be affected by empathy, and then, how empathic experiences change over time as a function of learning. This will bring us to the conclusion that vicarious emotional learning and empathy may be mutually reinforcing, and sometimes impossible to distinguish.

New Light on Old Phenomena

Despite many conceptual and empirical similarities, the study of vicarious learning and empathy has been confined to distinct research traditions, and limited cross talk between the two has resulted in little attention to common features. However, recent empirical work using neuroscientific and computational approaches has rediscovered old, and uncovered several new, links between the emotional learning underlying vicarious experiences and empathy. This development opens up new exciting avenues of research focusing on commonalities between the two phenomena, with potential consequences for both theory development and practical implementations.

Here, we identify two principal aspects of the relationship between empathy and emotional learning during vicarious experiences that have been described in the literature: (1) the impact of empathy on emotional learning, and (2) the ways learning influences empathy. These aspects should be viewed as nonexhaustive and simplified descriptions of the relevant research. Taken together, they show that empathic and learning processes are intimately interconnected, and may often be mutually reinforcing.

The first aspect describes how empathic processes impact individual learning. Research shows that basic emotional processes at the core of empathy, such as emotional contagion and resonance, convey information to the empathizer that updates his/her knowledge about the world. These processes can provide the basis for learning. Indeed, empathizing with another individual can be a fast and relatively safe way to infer the value of events and situations. For example, resonating, through social learning, with the disgust expressed by a person tasting a particular berry during a walk in the forest might save you from the potentially noxious experience. Of course, your empathic response might also bias learning about the situation in a nonadaptive way, if say, the person whose emotional experiences you are sharing is unreliable or your empathic response causes you to forego other, more important emotional information in the situation.

The second aspect of the relationship between empathy and emotional learning describes how empathy develops and changes through learning. For example, your empathic response to the disgust displayed by your berry eating friend might change over time as a function of new information, or by avoiding exposing yourself to her emotional displays. We will discuss ways in which learning can change empathic processes over time, and how this can be expressed in social situations. In fact, empathic experiences of an individual might themselves be a result of a long learning history.

As we will see, the different ways empathy and learning are related are also linked to situational factors and individual motives, as well as to stable behavioral dispositions, such as trait empathy. Based on the circumstances, empathy can intentionally or unintentionally change to conform to the situational demands or the observer's goals.

Due to space limitations, our discussion will primarily focus on aversive emotional processes in humans. The aversive valence spectrum contains some of the most obvious examples of links between emotional learning and empathy, yet this does not imply that the links between reward learning and empathic responses are less interesting. In fact, as will be discussed below, recent research has demonstrated several important similarities between self and other experienced reward in terms of their neural and behavioral principles (Burke, Tobler, Baddeley, & Schultz, 2010; Mobbs et al., 2009; Morelli et al., 2015; Schultz, Dayan, & Montague, 1997).

EMPATHY AFFECTS EMOTIONAL LEARNING

Your empathic response to the fearful person being harassed in a specific neighborhood or the apparently disgusted berry-picker in the examples above, might determine your memories of these events, as well as how your experiences will influence your future behavior. But how? The link between empathy and emotional learning is fundamentally related to their reliance on emotional states that are intrinsically motivating. As such, it has potentially important consequences for the self and/or for those who we care for. To understand how empathic processes affect emotional learning, we begin by specifying the meaning of empathy and identify the features of the emotional ingredients that is shared between empathy and vicarious learning.

Empathy is usually described as consisting of at least three components: emotional resonance (or experience sharing), perspective taking (or mentalizing), and motivated prosocial action (Baron-Cohen, 2005; Bernhardt & Singer, 2012; De Waal & Preston, 2017; FeldmanHall, Dalgleish, Evans, & Mobbs, 2015; Lamm, Decety, & Singer, 2011; Zaki, 2014; Zaki & Ochsner, 2012). For example, witnessing another individual in pain can elicit (1) an emotional resonance with the other's pain; (2) mentalizing about how the situation is experienced from the other's perspective; and finally (3) a motivation to act on it, for example, by removing the source of other's pain and/or providing comfort. These actions have been both measured and manipulated as state-dependent variables (Lamm et al., 2011; Zaki, 2014), and assessed in terms of stable behavioral traits (Davis, 1994; Mehrabian & Epstein, 1972). Depending on situational factors, empathic processes have also been shown to be either automatically displayed or related to the empathizer's motives and decision to empathize (Cameron, Spring, & Todd, 2017; Zaki, 2014). These observations illustrate the inherent malleability of empathy, which can change with learning. As will be discussed throughout this chapter, regardless of whether empathy is targeting pain, any other aversive affective states, or happiness, these processes share the emotional experience and the motivated behavioral aspects of vicarious learning. Empathy and vicarious learning recruit brain systems that are partly overlapping, and both can influence whether people choose to approach or avoid certain situations. To clarify this, we first turn the attention to some key concepts in learning theory, and then show how they are related to empathic processes.

Bridging Terminologies

Affect resonance is one of the key components of empathy. Such a response, for example, the agony of experiencing someone else's anxiety, or pleasure of sharing loved one's happiness, is identical to what learning theory refers to as unconditioned or a conditioned responses (UR or CR) elicited by the exposure to valued stimuli. These emotional responses constitute the basic building blocks of learning. An UR is elicited by an intrinsically aversive or rewarding unconditioned stimulus (US), and a CR is elicited by a conditioned stimulus (CS) that has acquired its value through previous pairing with an US. For example, the expression of pain in your new born child might be intrinsically aversive, and might cause more subtle cues that predict the infant's suffering to become CS. Pain expressions in your rival during a boxing match might, on the other hand, be rewarding because you have learned in the past that a rival's pain in this competitive social context predicts intrinsically rewarding outcomes to the self. In the latter case, the other's pain would constitute a CS, and your response a CR. Although the suffering of a competitor can serve as a motivating stimulus, its value is likely to differ greatly depending on the situation and the relationship between the observer and the competitor. The UR and CR people experience when exposed to others' emotions, thus seem to promote further learning.

Although direct (Pavlovian) conditioning focuses on associations between neutral and emotional stimuli (e.g., CS–US associations), instrumental learning examines how behavior is changed: strengthened or attenuated, by its emotional consequences. For example, related to the earlier example, watching your berry-picking friend's expression of disgust, might reinforce avoidance of the berries. If you would defy your friend's alarming facial cues and taste the berries, and find them delicious, your picking and eating behavior would be reinforced. If your expectations of a disgusting berry would be violated by a tasty experience, the memory of the event is strengthened. This way of learning about the behavioral consequences (increasing rewards and minimizing punishments) to one-self or others can be formalized by quantifying the underlying parameters using error-driven learning models. In brief, these models quantify the mismatch (i.e., prediction errors) between expected and actual reinforcements, leading to updated associations and changed behaviors. Motivated by the close match between neural activity during dopaminergic signaling and prediction error (Schultz et al., 1997), recent work on both vicarious learning and empathy has begun using formalized models to gain understanding of the underlying mechanisms of these two phenomena. For example, similar principles of error-driven learning seem to determine learning about emotional consequences to both self and others (Burke et al., 2010; Crockett, 2016; Joiner, Piva, Turrin, & Chang, 2017; Lindström et al., 2017), increased empathic responses (Hein, Engelmann, Vollberg, & Tobler, 2016), and the speed of learning as a function of empathic abilities (Lockwood, Apps, Valton, Viding, & Roiser, 2016). In sum, many aspects of both Pavlovian and instrumental learning theory can be directly related to empathic processes.

Emotional Resonance in Empathy

Accumulating evidence suggest that basic emotional resonance between individuals of the same species exists across mammals (De Waal & Preston, 2017; Jeon et al., 2010; Meyza, Bartal, Monfils, Panksepp, & Knapska, 2017). Findings have also demonstrated that such emotional expressions in rodents (Bartal, Decety, & Mason, 2011; Church, 1959; Langford et al., 2010) and chimpanzees (Masserman, Wechkin, & Terris, 1964) are linked to the motivation to help relieve conspecifics from pain and other aversive conditions across long periods of time. These findings taken together with the demonstration of similar empathic emotions that are linked to prosocial behavior in humans (Bernhardt & Singer, 2012; Lamm et al., 2011), strongly suggest a common phylogenetic origin (De Waal & Preston, 2017; Panksepp & Panksepp, 2013). These observations further support the claim that the emotional display of a conspecific can constitute valued stimuli akin to an US or a CS with the power to motivate behavior.

Much of the earlier research on empathic processes emphasized the automatic and uncontrollable character of dual representations in affective brain regions that instantly caused a matching emotional state (resonance) in the empathizer (Gallese, Keysers, & Rizzolatti, 2004). Yet, already in the 1980s, research showed that social context strongly affects socially "instigated" emotions. For example, Lanzetta and Englis (1989) demonstrated that the pain of a demonstrator believed to be a future competitor was not automatically shared by the observer. In fact, the emotional effect was the opposite, triggering a counter-mirroring response or "schadenfreude" (pleasure in another's pain). Similar effects accompanied with an attenuation of activity in the brain regions reliably implicated in the affective components of empathy: the anterior insula (AI) and the anterior cingulate cortex (ACC), were shown when the target person had previously been cheating in a simple economic game (Singer et al., 2006). Across species, social information about relatedness, similarity, and familiarity has also been shown to enhance the intensity of empathy and empathy-adjacent processes, including emotional resonance (Cameron et al., 2017b; Cialdini, Brown, Lewis, Luce, & Neuberg, 1997; D'Amato & Pavone, 1993; Lamm, Meltzoff, & Decety, 2010; Langford et al., 2006) and prosocial behavior (Bartal et al., 2011; Quervel-Chaumette et al., 2015). In humans, this effect has been demonstrated using phenotypical (e.g., ethnicity; Cikara, Bruneau, & Saxe, 2011; Hein et al., 2016; Xu, Zuo, Wang, & Han, 2009), experiential (e.g., pain experience; Cameron et al., 2017b; Lamm et al., 2010), and cultural (e.g., support for athletic team; Hein, Silani, Preuschoff, Batson, & Singer, 2010) similarities. Consistent with the view that the specific quality of these empathic responses serves as input to learning processes, the expression of learning and emotional memories would be shaped accordingly.

Recently, research has shown that emotional empathic responses are even more malleable than previously thought. Indeed, they are not only modulated by situational demands, but also by willful control in the observer (Cameron, Inzlicht, & Cunningham, 2017; Cameron & Payne, 2011; Keysers & Gazzola, 2014; Zaki, 2014). For example, one study showed that incarcerated psychopaths in fact evince physiological responses similar to normal controls when instructed to empathize, but not when empathizing spontaneously (Meffert, Gazzola, den Boer, Bartels, & Keysers, 2013). Another study applied multinomial modeling techniques to distinguish two types of emotional resonance—*intentional* and *unintentional* empathy—which differentially predict prosocial behavior (Cameron et al., 2017a). This suggests some empathic displays might be less of a question of empathy *ability* and more of empathy *propensity* (Keysers & Gazzola, 2014). These observations open up for the possibility to intervene and directly target learning processes that are dependent on emotional sharing. For example, if an individual has the capacity to emotionally resonate with others' emotional responses, yet lack the motivation to do so, learning interventions might be possible, for example, by establishing clear links between others' emotional response and intrinsically motivating experiences (US). Moreover, simple instructions might be used to direct attention to specific facial cues diagnostic for emotional expressions, as well as emotion-specific mental states, which in turn might initiate and facilitate learning.

Taken together, the studies reviewed here argue that resonance with another's emotional experience is the product of both relatively automatic, spontaneous sharing of another's emotional state, situational factors and relatively controlled, intentional resonance. The complexity of empathic processes is reflected in the temporal and spatial dynamic of different brain regions (Zaki & Ochsner, 2012). Research on the brain circuits implicated in empathy for pain includes core affective regions of the prefrontal (ACC) and somatosensory (AI) cortices. Moreover, the mentalizing aspects of empathy are thought to depend on the interplay between several cortical regions, among them the medial prefrontal cortex and the right temporal-parietal junction. In turn, these regions are thought to be supported by perceptually dependent input from the superior temporal sulcus, known to be important for the processing and understanding of biological movement (Adolphs, 2001), and the so-called mirror neuron system (encompassing both primary motor and parietal cortices; Gallese et al., 2004).

As we will see next, many of these brain regions are also highlighted in social forms of emotional learning. Importantly, their involvement is tied to similar behavioral and computational features in situations involving both empathy and learning. Despite this, the studies on humans reviewed so far, seldom, if ever, explicitly mention the possibility that empathic emotions contribute to learning, the formation of emotional memories, and reinforcement of behavior.

Vicarious Fear Learning

Others' emotional expressions do not only elicit empathic responses; they also change the observer's behavior through learning. For example, both human and animal studies on social fear learning provide ample evidence that emotional expressions of a conspecific (a "demonstrator") affect both the acquisition and learned attenuation (extinction) of threat responses (Colnaghi et al., 2016; Golkar & Olsson, 2016; Goubert, Vlaeyen, Crombez, & Craig, 2011; Gunnar, Hostinar, Sanchez, Tottenham, & Sullivan, 2015; Hooker, Germine, Knight, & D'Esposito, 2006; Kiyokawa, Hiroshima, Takeuchi, & Mori, 2014; Meffert et al., 2013; Olsson and Phelps, 2007). Animal work has shown that exposure to the sight, sound, or smell of a scared demonstrator triggers or potentiates threat responses

(Brechbühl, Klaey, & Broillet, 2008; Inagaki et al., 2014; Kim, Kim, Covey, & Kim, 2010; Lebowitz, Shic, Campbell, MacLeod, & Silverman, 2015), often referred to as fear contagion (Dezecache, Jacob, & Grèzes, 2015; Keum and Shin, 2016). The signaling of threat by the "demonstrator," when paired with a CS, may serve as a "social" US (analogous to pain stimuli to the self in direct fear conditioning), reinforcing the establishment of a learned threat response to the presented CS in the observer. For instance, in the previous example of vicarious transmitted disgust, your friend's facial expression when tasting the berry would serve as an US and the berry as the CS. Indeed, in humans and other primates, the social US is commonly represented by a facial expression of threat (Olsson & Phelps, 2007) or pain (Goubert et al., 2011). Classical studies on vicarious fear learning in monkeys have shown that upon observing the demonstrator's facial expressions of fear, the observer's own emotional response mimics the intensity of the demonstrator's expression, suggesting the same relationship as between an US and UR as in direct conditioning (Cook & Mineka, 1989).

Based on these findings in humans and other primates, it has been speculated that the potency of the social US is modulated by social information, such as empathic appraisals (Debiec & Olsson, 2017; Olsson & Phelps, 2007). This proposal has recently been more directly examined in a recent set of studies in humans. One of these (Lindström et al., 2017) submitted participants to both direct (Pavlovian) and vicarious (observational) fear learning. The results showed that the two kinds of learning were behaviorally indistinguishable, and could be explained by the same computational model for prediction-error learning. In addition, both direct and vicarious fear learning involved a cross-modal (self/other) aversive learning network, centered on the amygdala, the AI, and the ACC. Importantly, however, the information flow within, and the connectivity with, this aversive learning network differed in important ways between direct and social fear learning. Social fear learning was distinguished from direct fear learning by stronger connectivity with regions commonly implicated in perspective taking, primarily the TPJ. In addition, model comparisons showed that the input of information to the aversive learning network during vicarious learning (to the social US) came from the AI (and not from the amygdala as was the case during direct conditioning). Moreover, the strength of rated empathy with the demonstrator was positively related to activity in the AI to the US. Taken together, these findings further strengthened the claim that empathy plays a decisive role in the appraisal of the social US during vicarious emotional learning.

Also, the role of the ACC has been emphasized during vicarious fear learning in both humans and other animals (Olsson, Nearing, & Phelps, 2007). In support of this, studies in rodents have found that pharmacological inactivation of the ACC and optogenetic inhibition of the ACC–amygdala projections prevent vicarious fear learning, but leave direct fear learning intact (Allsop et al., 2016; Jeon et al., 2010; Jones & Monfils, 2016). This suggests a critical role of the ACC in vicarious learning, at least in some animals. Human and animal studies have also suggested the role of other pain processing sites in vicarious learning. For example, regions implicated by previous research in both nociceptive and empathic pain, such as the anterior insula (AI) and periaqueductal gray (PAG), have been implicated during vicarious learning (Chang & Debiec, 2016; Lindström et al., n.d.; Olsson et al., 2007). Moreover, blocking the endogenous opioid system, known to relieve self-experienced pain, enhances vicarious fear learning through changes in activity within the amygdala and the PAG (Haaker, Yi, Petrovic, & Olsson, 2017).

Although the findings reviewed so far are strongly suggestive, they are limited in terms of the causal conclusions that can be drawn about the tie between empathy and learning in humans. Addressing this concern directly, a recent study (Olsson et al., 2016) directly manipulated empathic appraisals before vicarious fear learning by means of a simple instruction manipulation that has been previously validated to enhance empathy with a target (encouraging the observer to pay close attention to a demonstrator's discomfort when receiving an aversive treatment; Batson, Early, & Salvarani, 1997). Receiving the empathy instruction (in comparison to receiving no instructions or instructions not related to empathy) augmented fear learning acquired through observing a demonstrator receiving electric shocks. In addition, this increase was especially noted among individuals displaying high trait empathy. These results are important because they provide evidence for a causal link between empathy and vicarious emotional learning. Interestingly, the relationship to trait empathy suggests that individuals high in this capacity are also able to utilize verbal instructions to a greater extent than individuals low in empathic capacity.

Vicarious Safety Learning

Interestingly, social cues can also serve as a potent safety signal. For example, the physical presence of a familiar conspecific may impair threat responses and fear learning following Pavlovian conditioning, which is referred to as social buffering of fear (Gunnar et al., 2015). Recently, it has been shown that in both humans and other primates, the presence of a calm and safe conspecific can immunize against direct fear learning (Golkar & Olsson, 2016; Mineka and Cook, 1986), as well as augment the extinction learning of already acquired threat responses (Golkar et al., 2013; Golkar, Castro, & Olsson, 2015). Interestingly, although these studies have not showed a direct relationship between safety learning and empathy with the demonstrator (presumably because the demonstrator did not express an emotional response to resonate with), they have showed that the strong extinction effect was dependent on the observer perceiving the demonstrator as safe (Golkar et al., 2013). Moreover, in accordance with the studies demonstrating a pro-ingroup bias in emotional sharing in empathy and vicarious fear learning, these studies have shown that vicarious safety learning is dependent on the similarity between the observer and the demonstrator (Golkar et al., 2015).

In this section, we have reviewed research on emotional processes in empathy, in particular, emotional sharing, suggesting that these processes can serve as UR and CRs (as characterized in the classical literature on Pavlovian and instrumental learning). This proposal is supported by recent research on vicarious learning by demonstrating that basic empathic processes can serve as input to the computations underlying social learning both through passive observation and reinforced avoidance behavior.

EMOTIONAL LEARNING AFFECTS EMPATHY

The research reviewed in the previous section strongly suggests that empathy (particularly emotional resonance/experience sharing) is integral to social emotional learning. By vicariously resonating with the emotional states of others, we can better learn the specific predictors of others' mental states, and then choose our approach and avoidance behavior accordingly. Our ability to learn others' emotions tracks with our ability to correctly perceive their mental states *qua* empathy, and varies with theoretically expected manipulations (e.g., is reduced when learning from outgroup members). But in turn, emotional learning can also influence empathy. In the mid to late 20th century, researchers discussed "vicariously instigated emotions" as URs paired with neutral cues (Berber, 1962; Hygge and Öhman, 1978; Lanzetta & Englis, 1989). Empathy was argued to be one such vicarious emotion, occurring when people experience congruent (or identical) emotions when witnessing those emotions in others (Lanzetta & Englis, 1989).

Learning of Empathy Across the Life Span

Theorists have speculated that empathy arises in part because of early childhood conditioning, in which children come to expect pleasurable experiences (e.g., cuddling) when caretakers express congruent affect. Over time, children's unconditioned empathic responses are paired with emotional expression in others and eventually occur in response to witnessing others' affective expression without need for direct reinforcement (Aronfreed, 1968; Hoffman, 1975, 1977). The importance of learning to develop empathic abilities is supported by the observation that, across species, individuals that were orphaned or socially deprived as infants show signs of deficits in empathic ability (Fries & Pollak, 2004). Yet, taken together, research in young humans and other animals indicates that the strong version of the learning account of empathy—that empathic processes are only shaped by learning—is unlikely to hold. Instead, primates and probably several mammals are likely to be born equipped with the basic neural machinery for the computation of empathic emotions (De Waal & Preston, 2017; Warneken & Tomasello, 2006).

Learning is not only important to the development of empathy in the young. Empathic expressions can be changed in adults as well. One study found that outgroup empathy was improved following a brief learning intervention. Receiving costly help from outgroup members elicited prediction errors in the anterior insular cortex, which predicted increased and generalized empathy for a different outgroup member (Hein et al., 2016). In addition, when individuals high in trait empathy engage in prosocial learning (i.e., learning to obtain rewards for others as opposed to themselves), the presence of a prediction error in the posterior subgenual ACC/basal forebrain selectively predicts successful and quicker prosocial learning (Lockwood et al., 2016). The link between prediction error and subsequent empathic/prosocial behavior is another example of learning influencing empathic outcomes.

Other research has demonstrated that empathy can be a function of an observer's previously learned associations between emotional expressions and reward/punishment outcomes. Specifically, one experiment found that observers empathized more when affective facial expressions were congruently paired with rewards or shocks, compared with incongruous pairings, and that these effects persisted into a direct test phase where emotional expressions were presented in the absence of reward or shock outcomes (Englis, Vaughan, and Lanzetta, 1982). In other words, facial expressions of emotion evoke empathy because of their previously learned predictive value.

Learned expectations about the costs and rewards associated with empathy also shift whether people choose to engage in empathy. For example, when anticipating a cooperative environment, people engage in more affective resonance with others' emotional expressions (Lanzetta & Englis, 1989), and incentivizing empathy increases empathy even among psychopaths and clinical narcissists (Arbuckle & Cunningham, 2012; Hepper, Hart, & Sedikides, 2014; Meffert et al., 2013). In addition, expecting empathy to be associated with costs (financial, emotional, etc.) predicts empathy avoidance (Andreoni, Rao, & Trachtman, 2011; Cameron, Harris, & Payne, 2016; Cameron & Payne, 2011; Pancer, McMullen, Kabatoff, Johnson, & Pond, 1979; Shaw, Batson, & Todd, 1994).

Emotional Learning and Empathy as Mutually Reinforcing Loops

The positive effects associated with empathy-motivated prosociality may lead to a feedback loop that in turn motivates further empathy and prosocial behavior. Vicariously sharing the aversive experience of a suffering target motivates an observer to approach the target to console them. This can reduce the negative affective state of both the target and the observer, whose personal distress is alleviated (Zaki, 2014). The reduction of the observer's negative affect is in itself a reward that may lead to empathy and prosociality. Further vicarious resonance with the target's relief and positive affect postconsolation—or simply the "warm glow" of having done a good deed (Andreoni et al., 2011)—will motivate the observer to exhibit empathy again in the future. Similar patterns are observed in animal models. For example, rats will spontaneously display helping behavior to free a trapped cage-mate. However, after being administered midazolam (an anxiolytic), this helping behavior was reduced, suggesting a role of affect resonance in motivating prosociality (Bartal et al., 2016). Rhesus macaques also demonstrate vicarious reinforcement when rewards are given to another monkey, and this vicarious reward motivates them to prefer cues paired with reward to the other monkey to those paired with reward to no one (Chang, Winecoff, & Platt, 2011). Future work should expand the aforementioned work on how cost/reward expectations predict empathic choice to investigate whether socially and vicariously learned cost/reward expectations for empathy predict empathy approach or avoidance.

A COMMON CORE

Vicarious resonance with others' emotional states—empathy—seems integral to the process of vicarious emotional learning. One must be able to vicariously resonate with another's emotional state to learn from it. Conditions which limit ability to empathize with a demonstrator (such as the demonstrator being a member of a social outgroup; Golkar et al., 2015—or in clinical psychopathy; Aniskiewicz, 1979) reduce vicarious learning ability. In contrast, encouraging empathy toward the demonstrator improves vicarious learning from that demonstrator (Olsson et al., 2016), as does trait empathy (Kleberg, Selbing, Lundqvist, Hofvander, & Olsson, 2015). In turn, emotional learning informs empathy. Emotional states are partly constructed from individual learning histories (Barrett, 2006) and empathy is no exception. In fact, empathy, in particular, seems subject to anticipated costs and rewards (Cameron et al., 2016; Zaki, 2014), with people selecting to engage in or avoid engaging in empathy in a motivated fashion that is partly based on empathy's anticipated outcomes.

Empathy and emotional learning seem to have a common evolutionary and biological basis, with both processes serving similar adaptive goals and undergirded by common neural networks. Across a lifespan, people use empathic processes to interpret others' emotional expressions to predict their environment (Klinnert, Campos, Sorce, Emde, & Svejda, 1983; Olsson & Ochsner, 2008). This enables them to predict and avoid dangerous situations, and to approach hedonically positive situations, which may be adaptive.

We suggest that empathy and emotional learning form a feedback loop. Empathizing with others is rewarding as it motivates prosocial behavior, which can induce positive affect both directly (through the "warm glow"), through vicarious resonance with the target's improved affect, and through social rewards for signaling empathy in societies where empathy is valued. Learning this reward structure encourages people to express empathy again in the future.

CONCLUSIONS AND FUTURE DIRECTIONS

In this chapter, we have argued that empathy and emotional learning can be mutually reinforcing, share common mechanisms and evolutionary incitements. To this end we have drawn upon the literature from animal models, human neuroscience, and human psychology. Others' emotional expressions induce both empathy and emotional learning, and empathic experiences also change over time due to continuous learning. Future work should investigate whether empathy can overwhelm social limits on vicarious emotional learning (such as group membership or demonstrator similarity to observer). In addition, the literature would benefit from studies tracking how people learn the cost/reward structure associated with empathy in an online fashion, as well as mapping the contributions of the so-called, model-free, model-based, and Pavlovian learning to the approach or avoidance resulting from empathy. Finally, by enhancing our understanding of the links between empathic and learning processes, we open up the possibility to study long-term effects of interventions to increase empathy and prosocial behavior.

Acknowledgments

We thank Daryl Cameron for comments on an earlier draft of this manuscript, and Tove Hensler for help with the manuscript. This work was supported by an independent Starting Grant (284366; Emotional Learning in Social Interaction) from the European Research Council, and the Knut and Alice Wallenberg Foundation (KAW 2014.0237) to A. Olsson, and by a National Science Foundation Graduate Research Fellowship and a National Science Foundation Graduate Research Opportunities Worldwide Fellowship to V. Spring.

References

Adolphs, R. (2001). The neurobiology of social cognition. *Current Opinion in Neurobiology*, 11(2), 231–239.

Allsop, S. A., Felix-Ortiz, A. C., Wichmann, R., Vienne, A., Beyeler, A., Nieh, E. H., & Tye, K. M. (2016). A cortico-amygdala circuit encodes observational fear learning. *Program no. 456.16/JJJ14. Neuroscience meeting planner*. San Diego, CA: Society for Neuroscience.

Andreoni, J., Rao, J. M., & Trachtman, H. (2011). Avoiding the ask: a field experiment on altruism, empathy, and charitable giving. Retrieved from http://www.nber.org/papers/w17648.

Aniskiewicz, A. S. (1979). Autonomic components of vicarious conditioning and psychopathy. *Journal of Clinical Psychology*, 35(1), 60–67.

Arbuckle, N. L., & Cunningham, W. A. (2012). Understanding everyday psychopathy: shared group identity leads to increased concern for others among undergraduates higher in psychopathy. *Social Cognition*, 30(5), 564–583.

Aronfreed, J. (1968). Conduct and conscience: The socialization of internalized control over behavior. *Review on Antimicrobial Resistance*. Oxford, England: Academic Press.

Bandura, A. (1979). Self-referent mechanisms in social learning theory. *American Psychologist*, 34, 439–441.

Baron-Cohen, S. (2005). Autism—"Autos": Literally, a Total Focus on the Self? In T. E. Feinberg, & J. P. Keenan (Eds.), *The lost self: pathologies of the brain and identity* (pp. 166–180). New York: Oxford University Press.

Barrett, L. F. (2006). Are emotions natural kinds. *Perspectives on Psychological Science*, 1(1), 28–58.

Bartal, I. B. -A., Decety, J., & Mason, P. (2011). Empathy and pro-social behavior in rats. *Science*, 334(6061), 1427–1430.

Batson, C. D., Early, S., & Salvarani, G. (1997). Perspective taking: imagining how another feels versus imaging how you would feel. *Personality and Social Psychology Bulletin*, 23(7), 751–758.

Ben-Ami Bartal, I., Shan, H., Molasky, N. M., Murray, T. M., Williams, J. Z., & Decety, J. (2016). Perspective taking: imagining how another feels versus imaging how you would feel. *Personality and Social Psychology Bulletin*, 8(7), 850.

Berber, S. M. (1962). Conditioning through vicarious instigation. *Psychological Review*, 69, 45–66.

Bernhardt, B. C., & Singer, T. (2012). The neural basis of empathy. *Annual Review of Neuroscience*, 35(1), 1–23.

Brechbühl, J., Klaey, M., & Broillet, M. -C. (2008). Grueneberg ganglion cells mediate alarm pheromone detection in mice. *Science*, 321(5892), 1092–1095.

Burke, C. J., Tobler, P. N., Baddeley, M., & Schultz, W. (2010). Neural mechanisms of observational learning. *Proceedings of the National Academy of Sciences of the United States of America*, 107(32), 14431–14436.

Cameron, C. D., Harris, L. T., & Payne, B. K. (2016). The emotional cost of humanity: anticipated exhausion motivates dehumanization of stigmatized targets. *Social Psychological and Personality Science*, 7(2), 105–112.

Cameron, D., Inzlicht, M., & Cunningham, W. (2017). Deconstructing empathy: a motivational framework for the apparent limits of empathy. Retrieved from https://osf.io/preprints/psyarxiv/d99bp/.

Cameron, C. D., & Payne, B. K. (2011). Escaping affect: how motivated emotion regulation creates insensitivity to mass suffering. *Journal of Personality and Social Psychology*, 100(1), 1–15.

Cameron, C. D., Spring, V. L., & Todd, A. R. (2017b). The empathy impulse: a multinomial model of intentional and unintentional empathy for pain. *Emotion*, 17(3), 395–411.

Chang, D. -J., & Debiec, J. (2016). Neural correlates of the mother-to-infant social transmission of fear. *Journal of Neuroscience Research*, 94(6), 526–534.

Chang, S. W. C., Winecoff, A. A., & Platt, M. L. (2011). Vicarious reinforcement in rhesus macaques (macaca mulatta). *Frontiers in Neuroscience*, 5, 27.

Church, R. M. (1959). Emotional reactions of rats to the pain of others. *Journal of Comparative and Physiological Psychology, 52*(2), 132–134.

Cialdini, R. B., Brown, S. L., Lewis, B. P., Luce, C., & Neuberg, S. L. (1997). Reinterpreting the empathy–altruism relationship: when one into one equals oneness. *Journal of Personality and Social Psychology, 73*(3), 481–494.

Cikara, M., Bruneau, E. G., & Saxe, R. R. (2011). Us and them: intergroup failures of empathy. *Current Directions in Psychological Science, 20*(3), 149–153.

Colnaghi, L., Clemenza, K., Groleau, S. E., Weiss, S., Snyder, A. M., Lopez-Rosas, M., & Levine, A. A. (2016). Social involvement modulates the response to novel and adverse life events in mice. *PLoS One, 11*(9), e0163077.

Cook, M., & Mineka, S. (1989). Observational conditioning of fear to fear-relevant versus fear-irrelevant stimuli in rhesus monkeys. *Journal of Abnormal Psychology, 98*(4), 448–459.

Crockett, M. (2016). How formal models can illuminate mechanisms of moral judgment and decision making. *Current Directions in Psychological Science, 25*, 85–90.

D'Amato, F. R., & Pavone, F. (1993). Endogenous opioids: a proximate reward mechanism for kin selection. *Behavioral and Neural Biology, 60*(1), 79–83.

Davis, M. (1994). Empathy. *A social psychological approach*. New York, NY: Westview Press.

Debiec, J., & Olsson, A. (2017). Social fear learning: from animal models to human function. *Trends in Cognitive Sciences, 21*(7), 546–555.

De Waal, F. B. M., & Preston, S. D. (2017). Mammalian empathy: behavioural manifestations and neural basis. *Nature Reviews Neuroscience, 18*(8), 498–509.

Dezecache, G., Jacob, P., & Grèzes, J. (2015). Emotional contagion: its scope and limits. *Trends in Cognitive Sciences, 19*(6), 297–299.

Englis, E. G., Vaughan, K. B., & Lanzetta, J. T. (1982). Conditioning of counterempathic emotional responses. *Journal of Experimental Social Psychology, 18*, 375–391.

FeldmanHall, O., Dalgleish, T., Evans, D., & Mobbs, D. (2015). Empathic concern drives costly altruism. *NeuroImage, 105*, 347–356.

Fries, A. B. W., & Pollak, S. D. (2004). Emotion understanding in postinstitutionalized Eastern European children. *Development and Psychopathology, 16*(2), 355–369.

Gallese, V., Keysers, C., & Rizzolatti, G. (2004). A unifying view of the basis of social cognition. *Trends in Cognitive Sciences, 8*(9), 396–403.

Golkar, A., Castro, V., & Olsson, A. (2015). Social learning of fear and safety is determined by the demonstrator's racial group. *Biology Letters, 11*(1), 20140817.

Golkar, A., & Olsson, A. (2016). Immunization against social fear learning. *Journal of Experimental Psychology: General, 145*(6), 665–671.

Golkar, A., Selbing, I., Flygare, O., Ohman, A., & Olsson, A. (2013). Other people as means to a safe end: vicarious extinction blocks the return of learned fear. *Psychological Science, 24*(11), 2182–2190.

Goubert, L., Vlaeyen, J. W. S., Crombez, G., & Craig, K. D. (2011). Learning about pain from others: an observational learning account. *Journal of Pain, 12*(2), 167–174.

Gunnar, M. R., Hostinar, C. E., Sanchez, M. M., Tottenham, N., & Sullivan, R. M. (2015). Parental buffering of fear and stress neurobiology: reviewing parallels across rodent, monkey, and human models. *Social Neuroscience, 10*(5), 474–478.

Haaker, J., Yi, J., Petrovic, P., & Olsson, A. (2017). Endogenous opioids regulate social threat learning in humans. *Nature Communications, 8* 15495.

Hein, G., Engelmann, J. B., Vollberg, M. C., & Tobler, P. N. (2016). How learning shapes the empathic brain. *Proceedings of the National Academy of Sciences of the United States of America, 113*(1), 80–85.

Hein, G., Silani, G., Preuschoff, K., Batson, C. D., & Singer, T. (2010). Neural responses to ingroup and outgroup members' suffering predict individual differences in costly helping. *Neuron, 68*(1), 149–160.

Hepper, E. G., Hart, C. M., & Sedikides, C. (2014). Moving narcissus: can narcissists be empathic? *Personality and Social Psychology Bulletin, 40*(9), 1079–1091.

Hoffman, M. L. (1975). Developmental synthesis of affect and cognition and its implications for altruistic motivation. *Developmental Psychology, 11*(5), 607–622 Empathy, its development and prosocial implications. Nebraska Symposium on Motivation 25, 169–217.

Hooker, C. I., Germine, L. T., Knight, R. T., & D'Esposito, M. (2006). Amygdala response to facial expressions reflects emotional learning. *Journal of Neuroscience, 26*(35), 8915–8922.

Hygge, S., & Öhman, A. (1978). Modeling processes in the acquisition of fears: vicarious electrodermal conditioning to fear-relevant stimuli. *Journal of Personality and Social Psychology, 36*, 271–279.

Inagaki, H., Kiyokawa, Y., Tamogami, S., Watanabe, H., Takeuchi, Y., & Mori, Y. (2014). Identification of a pheromone that increases anxiety in rats. *Proceedings of the National Academy of Sciences of the United States of America, 111*(52), 18751–18756.

Jeon, D., Kim, S., Chetana, M., Jo, D., Ruley, H. E., Lin, S. -Y., & Shin, H. -S. (2010). Observational fear learning involves affective pain system and Cav1. 2 Ca2+ channels in ACC. *Nature Neuroscience, 13*(4), 482–488.

Joiner, J., Piva, M., Turrin, C., & Chang, S. W. C. (2017). Social learning through prediction error in the brain. *NPJ Science of Learning, 2*(1), 8.

Jones, C. E., & Monfils, M. -H. (2016). Dominance status predicts social fear transmission in laboratory rats. *Animal Cognition, 19*(6), 1051–1069.

Keum, S., & Shin, H. S. (2016). Rodent models for studying empathy. *Neurobiology of Learning and Memory, 135*, 22–26.

Keysers, C., & Gazzola, V. (2014). Dissociating the ability and propensity for empathy. *Trends in Cognitive Sciences, 18*(4), 163–166.

Kim, E. J., Kim, E. S., Covey, E., & Kim, J. J. (2010). Social transmission of fear in rats: the role of 22-kHz ultrasonic distress vocalization. *PLoS One, 5*(12), e15077.

Kiyokawa, Y., Hiroshima, S., Takeuchi, Y., & Mori, Y. (2014). Social buffering reduces male rats' behavioral and corticosterone responses to a conditioned stimulus. *Hormones and Behavior, 65*(2), 114–118.

Kleberg, J. L., Selbing, I., Lundqvist, D., Hofvander, B., & Olsson, A. (2015). Spontaneous eye movements and trait empathy predict vicarious learning of fear. *International Journal of Psychophysiology, 98*(3), 577–583.

Klinnert, M. D., Campos, J. J., Sorce, J. F., Emde, R. N., & Svejda, M. (1983). Emotions as behavior regulators: social referencing in infancy. In R. Plutchnik, & H. Kellerman (Eds.), *The emotions* (pp. 57–86). New York: Academic Press Vol 2.

Laland, K. N. (2004). Social learning strategies. *Animal Learning & Behavior, 32*(1), 4–14.

Lamm, C., Decety, J., & Singer, T. (2011). Meta-analytic evidence for common and distinct neural networks associated with directly experienced pain and empathy for pain. *NeuroImage, 54*(3), 2492–2502.

Lamm, C., Meltzoff, A. N., & Decety, J. (2010). How do we empathize with someone who is not like us? A functional magnetic resonance imaging study. *Journal of Cognitive Neuroscience, 22*(2), 362–376.

Langford, D. J., Crager, S. E., Shehzad, Z., Smith, S. B., Sotocinal, S. G., Levenstadt, J. S., & Mogil, J. S. (2006). Social modulation of pain as evidence for empathy in mice. *Science, 312*(5782), 1967–1970.

Langford, D. J., Tuttle, A. H., Brown, K., Deschenes, S., Fischer, D. B., & Mutso, A. (2010). Social modulation of pain as evidence for empathy in mice. *Science, 5*(2), 163–170.

Lanzetta, J. T., & Englis, B. G. (1989). Expectations of cooperation and competition and their effects on observers' vicarious emotional responses. *Journal of Personality and Social Psychology, 56*(4), 543–554.

Lebowitz, E. R., Shic, F., Campbell, D., MacLeod, J., & Silverman, W. K. (2015). Avoidance moderates the association between mothers' and children's fears: findings from a novel motion-tracking behavioral assessment. *Depression and Anxiety, 32*(2), 91–98.

Lockwood, P. L., Apps, M. A. J., Valton, V., Viding, E., & Roiser, J. P. (2016). Neurocomputational mechanisms of prosocial learning and links to empathy. *Proceedings of the National Academy of Sciences, 113*(35), 9763–9768.

Lindström, B., Haaker, J., & Olsson, A. (2017). A common neural network differentially mediates direct and social fear learning. *Neuroimage, 167*, 121–129.

Masserman, J. H., Wechkin, S., & Terris, W. (1964). "Altruistic" behavior in rhesus monkeys. *American Journal of Psychiatry, 121*(6), 584–585.

Meffert, H., Gazzola, V., den Boer, J. A., Bartels, A. A. J., & Keysers, C. (2013). Reduced spontaneous but relatively normal deliberate vicarious representations in psychopathy. *Brain, 136*(8), 2550–2562.

Mehrabian, A., & Epstein, N. (1972). A measure of emotional empathy. *Journal of Personality, 40*(4), 525–543.

Meyza, K. Z., Bartal, I. B. -A., Monfils, M. H., Panksepp, J. B., & Knapska, E. (2017). The roots of empathy: through the lens of rodent models. *Neuroscience & Biobehavioral Reviews, 76*, 216–234.

Mineka, S., & Cook, M. (1986). Immunization against the observational conditioning of snake fear in rhesus monkeys. *Journal of abnormal psychology, 95*(4), 307–318.

Mobbs, D., Yu, R., Meyer, M., Passamonti, L., Seymour, B., Calder, A. J., & Dalgleish, T. (2009). A key role for similarity in vicarious reward. *Science, 324*(5929), 900.

Morelli, S. A., Lieberman, M. D., & Zaki, J. (2015). The emerging study of positive empathy. *Social and Personality Psychology Compass, 9*(2), 57–68.

Olsson, A., McMahon, K., Papenberg, G., Zaki, J., Bolger, N., & Ochsner, K. N. (2016). Vicarious fear learning depends on empathic appraisals and trait empathy. *Psychological Science, 27*(1), 25–33.

Olsson, A., Nearing, K. I., & Phelps, E. A. (2007). Learning fears by observing others: the neural systems of social fear transmission. *Social Cognitive and Affective Neuroscience, 2*(1), 3–11.

Olsson, A., & Ochsner, K. N. (2008). The role of social cognition in emotion. *Trends in Cognitive Sciences, 12*(2), 65–71.

Olsson, A., & Phelps, E. A. (2007). Social learning of fear. *Nature Neuroscience, 10*(9), 1095–1102.

Pancer, S. M., McMullen, L. M., Kabatoff, R. A., Johnson, K. G., & Pond, C. A. (1979). Conflict and avoidance in the helping situation. *Journal of Personality and Social Psychology, 37*(8), 1406–1411.

Panksepp, J., & Panksepp, J. B. (2013). Toward a cross-species understanding of empathy. *Trends in Neurosciences, 36*(8), 489–496.

Quervel-Chaumette, M., Dale, M., Marshall-Pescini, S., & Range, F. (2015). Familiarity affects other-regarding preferences in pet dogs. *Scientific Reports, 5*, 18102.

Schultz, W., Dayan, P., & Montague, P. R. (1997). A neural substrate of prediction and reward. *Science, 275*(5306), 1593–1599.

Shaw, L. L., Batson, C. D., & Todd, R. M. (1994). Empathy avoidance: forestalling feeling for another in order to escape the motivational consequences. *Journal of Personality and Social Psychology, 67*(5), 879–887.

Singer, P. (2011). *The expanding circle: ethics, evolution, and moral progress*. Princeton, NJ: Princeton University Press.

Singer, T., Seymour, B., O'Doherty, J. P., Stephan, K. E., Dolan, R. J., & Frith, C. D. (2006). Empathic neural responses are modulated by the perceived fairness of others. *Nature, 439*(7075), 466–469.

Warneken, F., & Tomasello, M. (2006). Altruistic helping in human infants and young chimpanzees. *Science, 311*(5765), 1301–1303.

Xu, X., Zuo, X., Wang, X., & Han, S. (2009). Do you feel my pain? Racial group membership modulates empathic neural responses. *Journal of Neuroscience, 29*(26), 8525–8529.

Zaki, J. (2014). Empathy: a motivated account. *Psychological Bulletin, 140*(6), 1608–1647.

Zaki, J., & Ochsner, K. (2012). The neuroscience of empathy: progress, pitfalls and promise. *Nature Neuroscience, 15*(5), 675–680.

3

The Neural Bases of Empathy in Humans

Claus Lamm, Livia Tomova
University of Vienna, Vienna, Austria

DEFINITIONS: EMPATHY AND RELATED TERMS

While numerous definitions exist (Batson, 2009), there is wide agreement that empathy is a multifaceted construct that involves intricate interactions of bottom-up and top-down components (Decety & Jackson, 2004; Singer & Lamm, 2009; Zaki & Ochsner, 2012, for review). Before discussing these components and their neural underpinnings in humans in more detail, we need to elaborate on definition issues that will allow us to distinguish empathy from concepts that are related to, yet distinct from the term. The distinction among emotion contagion; empathy, sympathy, and compassion; and prosocial and altruistic behaviors seems particularly vital in a book on empathy in nonhuman animals. In fact, much of previous reports on empathy in nonhuman animals pertains more to the domains of emotion contagion or prosocial behavior, rather than empathy in the sense that it is defined in the human social psychological and social neuroscience literature. In that literature, empathy is usually conceptualized as the ability to reexperience (to share) the feelings of another person, with full awareness that the other person is the source of one's affect (de Vignemont & Singer, 2006). This view allows a clear distinction from emotion contagion, which denotes the tendency to "catch" or "be contaged" by other people's emotions and has also been labeled as "primitive empathy" (Hatfield, Rapson, & Le, 2008). For example, babies start crying when they hear other babies cry, but they do so long before they develop a sense of self separate from others. On a conceptual level, emotion contagion can only contribute to the full-blown experience of empathy, as it lacks the defining component of self-awareness and self/other distinction. The necessity of self–other distinction has important practical implications, too, as without

Neuronal Correlates of Empathy. http://dx.doi.org/10.1016/B978-0-12-805397-3.00003-6

it, witnessing someone else's emotions might result in personal distress and a self-centered response in the observer (Decety & Lamm, 2011, for review). Moreover, definitions in the human literature stress that empathy can be triggered by not only direct observation, but also the imagination of the other's affective state. Taken together, this view proposes to regard emotion contagion as an important, yet distinct and neither necessary nor sufficient process for the experience of empathy. Putting self-awareness and self–other distinction as a requisite for "true, full-blown empathy" (in the sense the term is used in humans) does not imply, though, that nonhuman animals cannot experience empathy. While for quite some time many scholars would have agreed with such a conclusion, recent experimental evidence in, for example, ravens, as well as canines (Bugnyar, Reber, & Buckner, 2016; Muller, Schmitt, Barber, & Huber, 2015), two species known for their advanced social skills, suggest that we should not be so quick in dismissing animals as possessing skills that for a long time we have thought to be uniquely human.

Apart from emotion contagion and the related bottom-up, sensory-driven processes, we also need to distinguish empathy from higher-level and top-down regulated processes such as empathic concern, sympathy, or compassion (Singer & Klimecki, 2014, for review). These three terms usually describe experiences that not only encompass but also require sharing another's affect, in the sense of "feeling as" the other person. In addition, one also needs to "feel for" and care for that person, and become motivated to alleviate their distress or suffering. This distinction also implies that the "feeling for" aspect, when responding to the emotions of others resulting in prosocial and altruistic behaviors, that is, behavior that benefits others, such as helping or supporting them, usually is associated with a cost to oneself. This is not to say, though, that empathy (in the sense of "feeling as") plays no role in prosocial behavior and "feeling for" another person.—While distinct on a conceptual level, these phenomena are usually highly interrelated in their everyday occurrences. For instance emotion contagion triggered by seeing a portrait of a desperate refugee can result in empathy in the sense of sharing that person's affect, which in turn might trigger sympathy, and prompt prosocial actions (such as donating money or goods, or engagement in volunteer work). However, this example also serves to illustrate that prosocial action and the associated prosocial feelings and motivations are by no means an automatism. This can be shown by how many people respond now, after several years of an ongoing "refugee crisis," to very similar portraits—that is, with quite radically changed emotions and behaviors. From a scientific perspective, several factors can explain this shift in public and individual responses to similar stimuli. This includes bottom-up processes, such as habituation—seeing similar stimuli over and over again, which result in blunted affective responses. Top-down processes also play a role, such as when political messages playing with our fears result in refugees not being perceived any more as people in need, but as economic refugees who are threatening our welfare. Importantly, such modulations of empathy are additionally influenced by stress, an aspect that we will focus on in a later part of this chapter.

MIRROR NEURONS, SHARED REPRESENTATIONS, AND SELF–OTHER DISTINCTION

The fact that empathy "doesn't happen in a vacuum" brings us to the next part of this chapter, which is a brief review of how neuroscience research over the last 15 years has advanced our understanding of empathy and related phenomena.. The following three aspects deserve particular attention in the context of the present book: what is the role of "mirror neurons" in empathy; is affect sharing implemented by shared representations; and what is the role of self–other distinction?

Ever since the exciting discovery of mirror neurons in the early 90s (di Pellegrino, Fadiga, Fogassi, Gallese, & Rizzolatti, 1992), this special class of neurons, which respond both to executed and perceived actions, has been heavily linked to interpersonal phenomena. This went from simple processes, such as action coordination, to more elaborate ones, including empathy, up to high-level phenomena such as esthetic appreciation (Freedberg & Gallese, 2007). Notwithstanding recent criticism that some of these interpretations might have been overstretched (Hickok, 2009), mirror neurons certainly play an important role in many cases of empathy-like responses in nonhuman and human animals alike. It is, however, important to keep in mind that mirror neurons are neither a necessary nor a sufficient condition for empathy to arise (Lamm & Majdandzic, 2015, for review). This is so because empathy, as outlined previously, can also be evoked without direct action observation (e.g., by imagining how refugees described in a newspaper article without photos must feel like). Moreover, mirroring of actions does not necessarily lead to correct mirroring of the associated affective response—such as when being exposed to people whose sensorimotor–affective mappings differ from our own. This has been illustrated by experiments in which participants were exposed to situations that would be painful for themselves, but in reality were not painful for the observed other. The consistent finding of these studies was that mechanisms other than motor resonance alone mediate the empathic response and its regulation (Lamm, Meltzoff, & Decety, 2010; Lamm, Nusbaum, Meltzoff, & Decety, 2007). More specifically, areas involved in self–other distinction and cognitive control, such as the temporoparietal junction and prefrontal areas, were recruited when participants had to control their automatic yet inaccurate vicarious emotional responses (see also, for effects of stress on similar experimental manipulations). Hence, for the translation of these findings to the literature available on animals, it is important to keep in mind that conclusions such as that a species has mirror neurons is insufficient to show that this species also possesses empathy, or even emotion contagion.

Regarding the exact neural mechanisms enabling affect sharing, there is an ongoing debate whether this can be accounted for by so-called "shared representations" (Krishnan et al., 2016; Rütgen, Seidel, Riecansky, & Lamm, 2015; Rütgen, Seidel, Silani, et al., 2015; Zaki, Wager, Singer, Keysers, & Gazzola, 2016). The concept of shared representations builds up on the observation that empathy for a certain emotion recruits similar brain areas as the first-hand experience of that emotion. This has not only been observed very consistently, but also using a variety of neuroscience methods, such as functional magnetic resonance imaging (fMRI) (Lamm, Decety, & Singer, 2011), electroencephalography (EEG) (Bufalari, Aprile, Avenanti, Di Russo, & Aglioti, 2007), and transcranial magnetic stimulation (Avenanti, Bueti, Galati, & Aglioti, 2005). For instance fMRI research has consistently shown that parts of the network that are activated during self-experienced pain are also activated during empathy for pain, such as the mid-cingulate cortex and the anterior insular cortex (Lamm et al., 2011, for meta-analysis). The crucial question is what conclusions can be drawn from this activation overlap, as fMRI basically only provides information on "neural correlates," with limited spatial resolution. Hence, similarity in activation maps between two conditions (i.e., self- and other-related pain responses) might stem from different underlying neural populations, and hence different brain functions. While some recent evidence, using more refined multivariate fMRI analyses, suggests that this seems the case (Krishnan et al., 2016), psychopharmacological methods and variations in experimental design allowing more causal conclusions support a partially specific sharing of brain functions and representations (Rütgen, Seidel, Riecansky, et al., 2015; Rütgen, Seidel, Silani, et al., 2015). More specifically, these studies indirectly suggest an involvement of the opioid system in empathy for pain. Future studies will thus need to resolve this controversy, and to show whether affect sharing indeed relies on a specific sharing of affective experiences and their underlying neural computations, or whether previous neuroscience evidence only indicates rather domain-general processes such as vicarious arousal or engagement of the so-called salience network (though see Lamm, Silani, & Singer, 2015), in which sharing of pleasant affect activated areas outside the "salience network"). Resolving this debate is important not only for reasons of scientific curiosity and rigor, but also because of its implications for subsequent processes such as sympathy and prosocial behavior—as models relying on genuine and specific affect would make slightly different predictions as a more domain-general "salience account." Ultimately, the truth might very much lie in between the two explanations, with some parts of the activation overlap being specific, and others being domain-general.

The fact that empathy activates neural networks related to affective processing and regulation explains the necessity of self–other distinction, which, as outlined previously, is a defining feature of empathy. Recent progress in neuroimaging, lesion, and TMS research has tied this ability to a specific subregion in the human brain, the temporoparietal junction (Carter & Huettel, 2013; Donaldson, Rinehart, & Enticott, 2015; Lamm, Bukowski, & Silani, 2016, for review). In empathy, this area seems to be particularly important when disentangling one's own, often aversive, responses to the plight of another person, from the actual emotions experienced by that person. Being a multimodal association area, some scholars have even argued that this area might be at the root of seemingly unique human social-cognitive abilities (Saxe, 2006). Hence, with the recent advent of neuroimaging in a variety of nonhuman animal species, including dogs and birds (Andics, Gacsi, Farago, Kis, & Miklosi, 2014; De Groof et al., 2013), it will be very interesting to test such claims.

MODULATIONS OF EMPATHY—THE CASE OF STRESS

Recent research has shown that shared emotions of self and other can be profoundly modulated by situational factors. One ostensive example of this plasticity is shown by documentation of the effects of acute stress on empathy. Although this line of research is still in its infancy, there is evidence that acute stress might affect shared representations of self and other in profound ways. On a physiological level, stress is known to trigger adaptive responses in the body to reallocate resources [mainly modulated by the release of catecholamines and cortisol (Lovallo & Thomas, 2000; Sapolsky, Romero, & Munck, 2000)]. The physiological changes modulated by these adaptive responses have been shown to also profoundly affect cognition. Stress effects on cognition are particularly prevalent in attention (Elling et al., 2011; Vedhara, Hyde, Gilchrist, Tytherleigh, & Plummer, 2000), memory (Vedhara et al., 2000; Wolf, 2009) and decision making (Starcke & Brand, 2012). Crucially, the underlying level of experienced anxiety has been shown to modulate the effect of stress on cognitive functioning (Bishop, Duncan, Brett, & Lawrence, 2004; Hood, Pulvers, Spady, Kliebenstein, & Bachand, 2015). Research so far has shown that cognitive processes associated with "higher-order" cognition, such as strategic reasoning and self-control, seem to be impaired under stress (Maier, Makwana, & Hare, 2015; Starcke & Brand, 2012). However, other cognitive processes, such as focused attention and general alertness (Plessow, Fischer, Kirschbaum, & Goschke, 2011), learning and memory consolidation (Henckens, Hermans, Pu, Joels, & Fernandez, 2009), and working memory (Yuen et al., 2009, 2011), have been shown to improve under stress. This discrepancy has led to the view that stress selectively enhances cognitive and affective abilities that are beneficial for survival, at the expense of other mental abilities (Hermans et al., 2014).

However, the direction of the effects of stress on empathy and related phenomena is much less clear, as empirical research has shown both improvements and impairments in social interactions under stress. For example, several behavioral experiments have shown that acute stress and anxiety appear to decrease affect sharing (Buruck, Wendsche, Melzer, Strobel, & Dorfel, 2014; Martin et al., 2015; Negd, Mallan, & Lipp, 2011). Intriguingly, one of these contributions was able to show these effects in both humans and mice, with a surprisingly direct translation of magnitude and direction of the effects between species (Martin et al., 2015). However, other studies found the opposite, that is, stress can improve certain aspects of empathy. For example, in a behavioral study, it was shown that across different processing levels, self–other distinction was profoundly modulated during acute stress (Tomova, von Dawans, Heinrichs, Silani, & Lamm, 2014). However, the effects were modulated by sex/gender—while men showed decreased self–other distinction under stress, women displayed the opposite pattern, and increased their self–other distinction across different domains (Tomova et al., 2014). More specifically, stressed women showed reduced emotional egocentricity, improved cognitive perspective taking, and a reduction of automatic imitative tendencies. In contrast, men exposed to acute stress showed diminished self–other distinction on all three levels. Their emotional egocentricity bias increased, they showed lower perspective-taking abilities, and reduced ability to overcome automatic imitation tendencies. These results are in line with the tend-and-befriend hypothesis by Taylor and colleagues, which suggests that women show increased affiliative behavior when stressed (Taylor et al., 2000). As appropriate self–other distinction is a prerequisite for social skills such as empathy (Singer & Klimecki, 2014; Singer & Lamm, 2009) and perspective taking (Epley, Keysar, Van Boven, & Gilovich, 2004) , these findings might represent empirical evidence for the tend-and-befriend hypothesis.

Contradictory to these findings, several behavioral experiments with male participants also found increased prosocial behavior under stress (Buchanan & Preston, 2014; von Dawans, Fischbacher, Kirschbaum, Fehr, & Heinrichs, 2012; Margittai et al., 2015; Takahashi, Ikeda, & Hasegawa, 2007) and prosocial decision making under time pressure (Rand, Greene, & Nowak, 2012), indicating that automatic responses can produce prosocial outcomes in both, men and women. Viewed from a bioevolutionary point of view, these modulations in automatic reflexive responding to others' pain might represent adaptive modulations to increase prosociality, which overall might improve the survival of the group. However, it is not entirely clear how these findings of increased prosocial behavior under stress relate to evidence on decreased self–other distinction in men. A possible mechanism might have been identified in a recent fMRI study, which has shown stress-induced enhancements in magnitude of neural responding in the empathy for pain network in response to seeing someone else in pain (Tomova et al., 2016). This finding indicates that automatic bottom-up empathic responses to painful situations of others increase under stress. Thus a possible underlying mechanism might be that decreased self–other distinction leads to increased self–other merging, which ultimately leads to higher affect sharing when confronted with emotions of others. Intriguingly, these modulations have been shown in a male sample, and were also found to be associated with increased prosocial behavior.

Hence, these findings might provide further insights into the potential mechanisms of how stress, empathy, and prosocial behavior are connected. As increased self–other resonance has been proposed as an underlying mechanism driving prosocial behavior (Christov-Moore & Iacoboni, 2016); stress-induced decreases in self–other distinction (i.e., increased self–other resonance) might present an underlying mechanism of why individuals exhibit increased prosocial behavior under stress.

This does not yet explain, though, how women, who show increased self–other distinction under stress, could become more prosocial under stress. An alternative path, mainly based on a model proposed by Singer and Klimecki (2014), might be that increases in self–other distinction lead to increased compassion and, by this, increase prosocial behavior. Compassion, as defined by Singer and Klimecki, in contrast to empathy, would not mean sharing the suffering of the other, but is characterized by feelings of warmth, concern, and care for the other, as well as a strong motivation to improve the other's well being (Singer & Klimecki, 2014). Thus from this reasoning, both increased, as well as decreased self–other distinction, could lead to higher prosocial behavior, although based on different underlying motivations (i.e., affect sharing versus compassion).

OXYTOCIN AS A POTENTIAL MODERATOR OF STRESS EFFECTS ON EMPATHY?

It remains unclear though why exactly sex/gender differences appear to be so profound in the effects of stress on social emotions. One possible candidate to explain the effects might be the oxytocin system. The neuropeptide oxytocin is synthesized in the hypothalamus (Swaab, Pool, & Nijveldt, 1975; Vandesande & Dierickx, 1975) and then projected to the posterior pituitary, from which it is secreted into the systemic circulation (Brownstein, Russell, & Gainer, 1980). Oxytocin plays a crucial role not only in parturition and lactation (Gimpl & Fahrenholz, 2001), but also has been shown to function as a central regulator in social attachment and prosocial behaviors (Heinrichs, von Dawans, & Domes, 2009). For a number of reasons, oxytocin might be a strong candidate to explain these sex/gender differences on a physiological level (Heinrichs et al., 2009; Meyer-Lindenberg et al., 2011). First, women are known to show higher oxytocin release under stress than men (Carter, 2007; Jezova et al., 1996; Sanders et al., 1990). Furthermore, oxytocin has been shown to improve mind reading (Domes, Heinrichs, Michel, Berger, & Herpertz, 2007)—although see Lane et al. (2015)—and enhance emotional empathy (Hurlemann et al., 2010). Most importantly, two recent studies found that administration of oxytocin leads to sharpened self–other perception (Colonnello, Chen, Panksepp, & Heinrichs, 2013) and improved self–other distinction on a cognitive level (Tomova et al., 2014). Thus the gender differences in self–other distinction under stress might be explained by gender differences in oxytocin release under stress, and the neuropeptide's positive effects on coping with stress (Heinrichs, Baumgartner, Kirschbaum, & Ehlert, 2003).

Crucially, though, more empirical evidence is needed to support these ideas and predictions. For example, the proposed mechanism suggesting increased compassion rather than affect sharing under stress in women has not been empirically supported so far. Due to the fact that women show different cortisol responding to stressors depending on their menstrual cycle phase and hormonal contraceptive usage (Kirschbaum, Kudielka, Gaab, Schommer, & Hellhammer, 1999), most stress studies exclude women from participation. Thus the empirical evidence of increased prosociality under stress so far is restricted to male participants (although, ironically, the original tend-and-befriend hypothesis stated these increases in prosociality to be a specifically female stress response (Taylor et al., 2000)). Therefore a very important next step in this line of research is to empirically investigate prosocial behavior under stress in women. In addition, future empirical research should address whether women in fact show increased compassion under stress, which would suggest that there is in fact a dissociation in the underlying mechanisms of how stress affects prosociality between the two genders.

In conclusion, future research on empathy should consider the factors sex/gender and stress, as most neuroscience experiments in humans and nonhuman animals involve stressful procedures whose effects should thus be taken into account when interpreting such research. Furthermore, adding endocrinological measures or experimentally manipulating them in neuroscience research might enable a more mechanistic understanding of these modulations, and the resulting sex/gender differences.

References

Andics, A., Gacsi, M., Farago, T., Kis, A., & Miklosi, A. (2014). Voice-sensitive regions in the dog and human brain are revealed by comparative FMRI. *Current Biology, 24*(5), 574–578.

Avenanti, A., Bueti, D., Galati, G., & Aglioti, S. M. (2005). Transcranial magnetic stimulation highlights the sensorimotor side of empathy for pain. *Nature Neuroscience, 8*(7), 955–960.

Batson, C. D. (2009). These things called empathy. In J. Decety, & W. Ickes (Eds.), *The social neuroscience of empathy* (pp. 3–16). Cambridge, MA: The MIT Press.

Ben-Ami Bartal, I., Decety, J., & Mason, P. (2011). Empathy and pro-social behavior in rats. *Science, 334*(6061), 1427–1430.

Bishop, S., Duncan, J., Brett, M., & Lawrence, A. D. (2004). Prefrontal cortical function and anxiety: Controlling attention to threat-related stimuli. *Nature Neuroscience, 7*(2), 184–188.

Brownstein, M. J., Russell, J. T., & Gainer, H. (1980). Synthesis, transport, and release of posterior pituitary hormones. *Science, 207*(4429), 373–378.

Buchanan, T. W., & Preston, S. D. (2014). Stress leads to prosocial action in immediate need situations. *Frontiers in Behavioral Neuroscience, 8,* 5.

Bufalari, I., Aprile, T., Avenanti, A., Di Russo, F., & Aglioti, S. M. (2007). Empathy for pain and touch in the human somatosensory cortex. *Cerebral Cortex, 17*(11), 2553–2561.

Bugnyar, T., Reber, S. A., & Buckner, C. (2016). Ravens attribute visual access to unseen competitors. *Nature Communications, 7,* 10506.

Buruck, G., Wendsche, J., Melzer, M., Strobel, A., & Dorfel, D. (2014). Acute psychosocial stress and emotion regulation skills modulate empathic reactions to pain in others. *Frontiers in Psychology, 5,* 517.

Carter, C. S. (2007). Sex differences in oxytocin and vasopressin: implications for autism spectrum disorders? *Behavioral Brain Research, 176*(1), 170–186.

Carter, R. M., & Huettel, S. A. (2013). A nexus model of the temporal-parietal junction. *Trends in Cognitive Sciences, 17*(7), 328–336.

Colonnello, V., Chen, F. S., Panksepp, J., & Heinrichs, M. (2013). Oxytocin sharpens self-other perceptual boundary. *Psychoneuroendocrinology, 38*(12), 2996–3002.

Christov-Moore, L., & Iacoboni, M. (2016). Self-other resonance, its control and prosocial inclinations: brain-behavior relationships. *Human Brain Mapping, 37*(4), 1544–1558.

von Dawans, B., Fischbacher, U., Kirschbaum, C., Fehr, E., & Heinrichs, M. (2012). The social dimension of stress reactivity: Acute stress increases prosocial behavior in humans. *Psychological Science, 23*(6), 651–660.

De Groof, G., Jonckers, E., Gunturkun, O., Denolf, P., Van Auderkerke, J., & Van der Linden, A. (2013). Functional MRI and functional connectivity of the visual system of awake pigeons. *Behavioural Brain Research, 239,* 43–50.

Decety, J., & Jackson, P. L. (2004). The functional architecture of human empathy. *Behavioral and Cognitive Neuroscience Reviews, 3,* 71–100.

Decety, J., & Lamm, C. (2011). Empathy vs. personal distress. In J. Decety, & W. Ickes (Eds.), *The social neuroscience of empathy* (pp. 199–214). Cambridge: MIT Press.

Domes, G., Heinrichs, M., Michel, A., Berger, C., & Herpertz, S. C. (2007). Oxytocin improves "mind-reading" in humans. *Biological Psychiatry, 61*(6), 731–733.

Donaldson, P. H., Rinehart, N. J., & Enticott, P. G. (2015). Noninvasive stimulation of the temporoparietal junction: A systematic review. *Neuroscience and Biobehavioral Reviews, 55,* 547–572.

Elling, L., Steinberg, C., Bröckelmann, A. K., Dobel, C., Bölte, J., & Junghofer, M. (2011). Acute stress alters auditory selective attention in humans independent of HPA: A study of evoked potentials. *PLoS One, 6*(4), .

Epley, N., Keysar, B., Van Boven, L., & Gilovich, T. (2004). Perspective taking as egocentric anchoring and adjustment. *Journal of Personality and Social Psychology, 87*(3), 327–339.

Freedberg, D., & Gallese, V. (2007). Motion, emotion and empathy in esthetic experience. *Trends in Cognitive Sciences, 11*(5), 197–203.

Gimpl, G., & Fahrenholz, F. (2001). The oxytocin receptor system: Structure, function, and regulation. *Physiological Reviews, 81*(2), 629–683.

Hatfield, E., Rapson, R. L., & Le, Y. L. (2008). Emotional contagion and empathy. In J. Decety, & W. Ickes (Eds.), *The social neuroscience of empathy* (pp. 19–30). Boston, MA: MIT Press.

Heinrichs, M., Baumgartner, T., Kirschbaum, C., & Ehlert, U. (2003). Social support and oxytocin interact to suppress cortisol and subjective responses to psychosocial stress. *Biological Psychiatry, 54*(12), 1389–1398.

Heinrichs, M., von Dawans, B., & Domes, G. (2009). Oxytocin, vasopressin, and human social behavior. *Frontiers in Neuroendocrinology, 30*(4), 548–557.

Henckens, M. J. A. G., Hermans, E. J., Pu, Z. W., Joels, M., & Fernandez, G. N. (2009). Stressed memories: How acute stress affects memory formation in humans. *Journal of Neuroscience, 29*(32), 10111–10119.

Hermans, E. J., Henckens, M. J., Joels, M., & Fernandez, G. (2014). Dynamic adaptation of large-scale brain networks in response to acute stressors. *Trends in Neuroscience, 37*(6), 304–314.

Hickok, G. (2009). Eight problems for the mirror neuron theory of action understanding in monkeys and humans. *Journal of Cognitive Neuroscience, 21*(7), 1229–1243.

Hood, A., Pulvers, K., Spady, T. J., Kliebenstein, A., & Bachand, J. (2015). Anxiety mediates the effect of acute stress on working memory performance when cortisol levels are high: A moderated mediation analysis. *Anxiety, Stress, and Coping, 28*(5), 545–562.

Hurlemann, R., Patin, A., Onur, O. A., Cohen, M. X., Baumgartner, T., Metzler, S., et al. (2010). Oxytocin enhances amygdala-dependent, socially reinforced learning and emotional empathy in humans. *Journal of Neuroscience, 30*(14), 4999–5007.

Jezova, D., Jurankova, E., Mosnarova, A., Kriska, M., & Skultetyova, I. (1996). Neuroendocrine response during stress with relation to gender differences. *Acta Neurobiologiae Experimentalis, 56*(3), 779–785.

Kirschbaum, C., Kudielka, B. M., Gaab, J., Schommer, N. C., & Hellhammer, D. H. (1999). Impact of gender, menstrual cycle phase, and oral contraceptives on the activity of the hypothalamus-pituitary-adrenal axis. *Psychosomatic Medicine, 61*(2), 154–162.

Krishnan, A., Woo, C. W., Chang, L. J., Ruzic, L., Gu, X., Lopez-Sola, M., et al. (2016). Somatic and vicarious pain are represented by dissociable multivariate brain patterns. *eLife, 5.*

Lamm, C., Bukowski, H., & Silani, G. (2016). From shared to distinct self-other representations in empathy: Evidence from neurotypical function and socio-cognitive disorders. *Philosophical Transactions of the Royal Society of London. Series B, Biological Sciences, 371*(1686), 20150083.

Lamm, C., Decety, J., & Singer, T. (2011). Meta-analytic evidence for common and distinct neural networks associated with directly experienced pain and empathy for pain. *NeuroImage, 54*(3), 2492–2502.

Lamm, C., & Majdandzic, J. (2015). The role of shared neural activations, mirror neurons, and morality in empathy--a critical comment. *Neuroscience Research, 90*, 15–24.

Lamm, C., Meltzoff, A. N., & Decety, J. (2010). How do we empathize with someone who is not like us? A functional magnetic resonance imaging study. *Journal of Cognitive Neuroscience, 22*(2), 362–376.

Lamm, C., Nusbaum, H. C., Meltzoff, A. N., & Decety, J. (2007). What are you feeling? Using functional magnetic resonance imaging to assess the modulation of sensory and affective responses during empathy for pain. *PLoS One, 12*, e1292.

Lamm, C., Silani, G., & Singer, T. (2015). Distinct neural networks underlying empathy for pleasant and unpleasant touch. *Cortex, 70*, 79–89.

Lane, A., Mikolajczak, M., Treinen, E., Samson, D., Corneille, O., de Timary, P., et al. (2015). Failed replication of oxytocin effects on trust: The envelope task case. *PLoS One, 10*(9), .

Lovallo, W. R., & Thomas, T. L. (2000). Stress hormones in psychophysiological research: Emotional, behavioral, and cognitive implications. In J. Cacioppo, L. G. Tassinary, & G. G. Berntson (Eds.), *Handbook of psychophysiology* (2nd ed., pp. 342–367). New York: Cambridge University Press.

Maier, S. U., Makwana, A. B., & Hare, T. A. (2015). Acute stress impairs self-control in goal-directed choice by altering multiple functional connections within the brain's decision circuits. *Neuron, 87*(3), 621–631.

Margittai, Z., Strombach, T., van Wingerden, M., Joels, M., Schwabe, L., & Kalenscher, T. (2015). A friend in need: Time-dependent effects of stress on social discounting in men. *Hormones and Behavior, 73*, 75–82.

Martin, L. J., Hathaway, G., Isbester, K., Mirali, S., Acland, E. L., Niederstrasser, N., et al. (2015). Reducing social stress elicits emotional contagion of pain in mouse and human strangers. *Current Biology, 25*(3), 326–332.

Meyer-Lindenberg, A., Domes, G., Kirsch, P., & Heinrichs, M. (2011). Oxytocin and vasopressin in the human brain: social neuropeptides from translational medicine. *Nature Reviews Neuroscience, 12*(9), 524–538.

Muller, C. A., Schmitt, K., Barber, A. L., & Huber, L. (2015). Dogs can discriminate emotional expressions of human faces. *Current Biology, 25*(5), 601–605.

Negd, M., Mallan, K. M., & Lipp, O. V. (2011). The role of anxiety and perspective-taking strategy on affective empathic responses. *Behaviour Research and Therapy, 49*(12), 852–857.

di Pellegrino, G., Fadiga, L., Fogassi, L., Gallese, V., & Rizzolatti, G. (1992). Understanding motor events: A neurophysiological study. *Experimental Brain Research, 91*(1), 176–180.

Plessow, F., Fischer, R., Kirschbaum, C., & Goschke, T. (2011). Inflexibly focused under stress: Acute psychosocial stress increases shielding of action goals at the expense of reduced cognitive flexibility with increasing time lag to the stressor. *Journal of Cognitive Neuroscience, 23*(11), 3218–3227.

Rand, D. G., Greene, J. D., & Nowak, M. A. (2012). Spontaneous giving and calculated greed. *Nature, 489*(7416), 427–430.

Rütgen, M., Seidel, E. M., Riecansky, I., & Lamm, C. (2015a). Reduction of empathy for pain by placebo analgesia suggests functional equivalence of empathy and first-hand emotion experience. *Journal of Neuroscience, 35*(23), 8938–8947.

Rütgen, M., Seidel, E. M., Silani, G., Riecansky, I., Hummer, A., Windischberger, C., et al. (2015b). Placebo analgesia and its opioidergic regulation suggest that empathy for pain is grounded in self pain. *Proceedings of the National Academy of Sciences of the United States of America, 112*(41), 5638–5646.

Sanders, G., Freilicher, J., Lightman, S. L., & Heinrichs, M. (1990). Psychological stress exposure to uncontrollable noise increases plasma oxytocin in high emotionality women. *Psychoneuroendocrinology, 15*(1), 47–58.

Sapolsky, R. M., Romero, L. M., & Munck, A. U. (2000). How do glucocorticoids influence stress responses? Integrating permissive, suppressive, stimulatory, and preparative actions. *Endocrinological Review, 21*(1), 55–89.

Saxe, R. (2006). Uniquely human social cognition. *Current Opinion in Neurobiology, 16*(2), 235–239.

Singer, T., & Klimecki, O. M. (2014). Empathy and compassion. *Current Biology, 24*(18), R875–R878.

Singer, T., & Lamm, C. (2009). The social neuroscience of empathy. *Annals of the New York Academy of Sciencess, 1156*, 81–96.

Starcke, K., & Brand, M. (2012). Decision making under stress: A selective review. *Neuroscience and Biobehavioral Reviews, 36*(4), 1228–1248.

Swaab, D. F., Pool, C. W., & Nijveldt, F. (1975). Immunofluorescence of vasopressin and oxytocin in the rat hypothalamo-neurohypophypopseal system. *Journal of Neural Transmission, 36*(3–4), 195–215.

Takahashi, T., Ikeda, K., & Hasegawa, T. (2007). Social evaluation-induced amylase elevation and economic decision-making in the dictator game in humans. *Neuroendocrinology Letters, 28*(5), 662–665.

Taylor, S. E., Klein, L. C., Lewis, B. P., Gruenewald, T. L., Gurung, R. A., & Updegraff, J. A. (2000). Biobehavioral responses to stress in females: Tend-and-befriend, not fight-or-flight. *Psychological Review, 107*(3), 411–429.

Tomova, L., von Dawans, B., Heinrichs, M., Silani, G., & Lamm, C. (2014). Is stress affecting our ability to tune into others? Evidence for gender differences in the effects of stress on self-other distinction. *Psychoneuroendocrinology, 43*, 95–104.

Tomova, L., Majdandžić, J., Hummer, A., Windischberger, C., Heinrichs, M., & Lamm, C. (2016). Increased neural responses to empathy for pain might explain how acute stress increases prosociality. *Social Cognitive and Affective Neuroscience* EPub ahead of print.

Vandesande, F., & Dierickx, K. (1975). Identification of the vasopressin producing and of the oxytocin producing neurons in the hypothalamic magnocellular neurosecretroy system of the rat. *Cell and Tissue Research, 164*(2), 153–162.

Vedhara, K., Hyde, J., Gilchrist, I. D., Tytherleigh, M., & Plummer, S. (2000). Acute stress, memory, attention and cortisol. *Psychoneuroendocrinology, 25*(6), 535–549.

de Vignemont, F., & Singer, T. (2006). The empathic bra how when why? *Trends in Cognitive Sciences, 10*(10), 435–441.

Wolf, O. T. (2009). Stress and memory in humans: Twelve years of progress? *Brain Research, 1293*, 142–154.

Yuen, E. Y., Liu, W., Karatsoreos, I. N., Feng, J., McEwen, B. S., & Yan, Z. (2009). Acute stress enhances glutamatergic transmission in prefrontal cortex and facilitates working memory. *Proceedings of the National Academy of Sciences of the United States of America, 106*(33), 14075–14079.

Yuen, E. Y., Liu, W., Karatsoreos, I. N., Ren, Y., Feng, J., McEwen, B. S., et al. (2011). Mechanisms for acute stress-induced enhancement of glutamatergic transmission and working memory. *Molecular Psychiatry, 16*(2), 156–170.

Zaki, J., & Ochsner, K. N. (2012). The neuroscience of empathy: Progress, pitfalls and promise. *Nature Neuroscience, 15*(5), 675–680.

Zaki, J., Wager, T. D., Singer, T., Keysers, C., & Gazzola, V. (2016). The anatomy of suffering: Understanding the relationship between nociceptive and empathic pain. *Trends in Cognitive Sciences, 20*(4), 249–259.

Neural Correlates of Empathy in Humans, and the Need for Animal Models

Christian Keysers*,**, Valeria Gazzola*,**

*Netherlands Institute for Neuroscience, an Institute of the Royal Netherlands Academy of Art and Sciences (KNAW), Amsterdam, The Netherlands; **University of Amsterdam, Amsterdam, The Netherlands

EMPATHY AND ITS COMPONENTS: EMOTIONAL CONTAGION, EMPATHY PROPER, AND SYMPATHY

Empathy often refers to a family of related processes through which we intuitively relate to the inner states of others. Although definitions vary, there is some consensus that one should discriminate at least three different phenomena when it comes to how an individual (human or animal) reacts to the emotions of another individual (Keysers, 2011; de Vignemont & Singer, 2006; Wispé, 1986). We will refer throughout this text to the individual experiencing empathy as a "witness" (rather than an observer, to underline the importance of senses other than vision), and to the person that is the object of the empathy as the "object." The first and simplest phenomenon is often referred to as *emotional contagion*, and occurs whenever a witness's emotional states come to resemble those of the object. For this simplest form to occur, it is not necessary for the witness to clearly attribute her triggered internal feelings to the feelings of the object. A prototypical example is when a newborn baby starts crying when witnessing other babies cry. The second phenomenon, *empathy proper*, occurs when the witness experiences emotional states that resemble those of the object and is aware that these states

Neuronal Correlates of Empathy. http://dx.doi.org/10.1016/B978-0-12-805397-3.00004-8

are experienced on behalf of the object. In both emotional contagion and empathy, a witness' triggered emotional states thus resemble those of the object of empathy, such that witnessing a person demonstrating distress triggers distress in the witness. This is where the third phenomenon, *sympathy*, differs: sympathy occurs when the witness no longer directly experiences emotions resembling those of the object, but rather prosocial emotions that are appropriate to help the object. Sympathy for a person in distress is then no longer distress, but a combination of distress and a "warm-hearted feeling" that motivates the witness to console and help the object.

It is generally believed that these three phenomena are causally related, such that overt signs of an object's emotions first trigger that emotion in a witness through emotional contagion. If the meta-cognitive apparatus of the witness enables her/him to do so, this contaged emotion is then attributed to the object of empathy, triggering an additional layer of empathy proper. If the witness is so disposed, this can then lead to the prosocial motivation called sympathy.

This trilayered distinction finds its routes in human psychology, where these phenomena are thought to appear in succession over the course of development, with emotional contagion being present soon after birth, while empathy proper and sympathy appear later, when children develop the ability to represent the self and others separately, and are able to regulate their own emotions well enough to help others (Decety & Jackson, 2004). Directly applying these distinctions in a cross-species approach is challenging: although rodents show signs of altering their own emotions in reaction to those of others (Atsak et al., 2011; Panksepp & Panksepp, 2013) and also demonstrate a motivation to help animals in needs (Ben-Ami Bartal, Decety, & Mason, 2011; Greene, 1969; Rice & Gainer, 1962), it is not obvious how one would establish whether rats are aware that the emotion they experience is that of another rat. Whether they experience emotional contagion, empathy proper, or sympathy thus remains difficult to assess. We therefore recommend using the term empathy and study its neural basis initially in its widest sense to gain traction on how the emotions of an object affect those of the witness, and then later attempt to distinguish emotional contagion, empathy proper, and sympathy.

These three layers of social transmission of affect exist in the context of a number of related phenomena. Some of them relate to how we perceive the actions of others, as well as their somatosensory states. Some are related to more explicit cognitive processes and have been referred to as mentalizing. Before focusing on emotional empathy, we will therefore briefly touch on this context.

RELATED PHENOMENA: MENTALIZING, SOMATOSENSORY, AND MOTOR EMPATHY

Mentalizing refers to people's ability to consciously think about the beliefs, desires, and intensions of others. Mentalizing is sometimes also referred to as "cognitive empathy." If we see a friend in great distress, and start to mentally deliberate that this is probably because he did not know that an important grant deadline had been moved, we engage in mentalizing. If the same friend simply makes us feel agitated without us engaging in the attribution of specific beliefs, we experience emotional contagion or empathy (depending on how much we attribute our distress to him). Clearly, these phenomena can—and often will—complement each other. However, because of space constrains, we will not be able to touch explicitly on the relation between the neural correlates of mentalizing and other forms of empathy, but a discussion of this relation can be found in Keysers and Gazzola (2007).

Somatosensory and motor empathy, on the other hand, refers to the fact that while witnessing other people's movements, we can feel ourselves into their movements as if we were moving in similar ways, a phenomenon sometimes also called kinesthetic empathy, particularly in the scientific study of the esthetics of dance (Jola, Ehrenberg, & Reynolds, 2012). Because we cannot telepathically feel the emotions of others, perceiving them must be based on the object's behavior: her bodily movements, facial expressions, vocalizations, the activity of glands (tears, sweat, pheromones) and, in humans, linguistic behavior—or the conspicuous lack of any of those. As such, mechanisms that allow us to process the actions of others are highly relevant to processing their emotions, which is why we will start by reviewing what we know about the neural correlates of kinesthetic empathy before addressing the neural correlates of emotional contagion in humans.

Our understanding of how the brain processes the actions of others has been strongly influenced by the discovery of mirror neurons in the macaque monkey (Gallese, Fadiga, Fogassi, & Rizzolatti, 1996). These motor neurons, first described in the ventral premotor cortex of the macaque monkey, respond both when a monkey performs a goal-directed action, for instance breaking the shell of a peanut, and when witnessing a similar action performed by others, be it by only seeing someone performing a silent version of this action or by only hearing the cracking of the shell or by hearing and seeing the action (Keysers et al., 2003; Kohler et al., 2002). Because electrostimulating the premotor cortex of the monkey triggers the performance of similar actions (Graziano, 2016), this suggest that witnessing the actions of others triggers a motor representation fit for performing a similar action in the witness. About 10% of the motor neurons in the ventral premotor cortex of the macaque have this mirror property, while the remaining 90% do not respond to the sight or sound of actions. Because the firing of premotor neurons is associated not with simple movements of muscles, but with complex, goal-directed actions such as grasping or tearing, and often do so independently of whether the action is performed with the right or the left hand, or even with the mouth (Gallese et al., 1996), this suggests that what is encoded in these neurons is not simple movements, such as flexing the biceps, but a higher-level action policy of what the action is to achieve (e.g., cracking a peanut's shell). This goal-directed coding becomes particularly apparent in mirror neurons that respond to the sound of actions: if a neuron responds when the monkey himself cracks a peanut, as well as when listening to the sound of a cracking peanut shell, what is common is the goal (cracking the shell), not the means—as the body movements achieving the goal are not apparent from the sound (Keysers, 2011).

A decade later, similar neurons were reported in the rostral inferior parietal cortex (Fogassi et al., 2005) of the monkey, a region that has direct connections to the ventral premotor cortex, which is also involved in the motor control of goal-directed actions.

In humans, rigorous functional magnetic resonance imaging studies have mapped all brain regions that have voxels that are activated both while executing and witnessing the actions of others. For instance we had participants view a number of complex goal-directed hand actions and execute such actions while their brain activity was measured using fMRI (Gazzola & Keysers, 2009). Voxels were then identified as being shared between action observation and execution, and hence to be potentially involved in kinesthetic empathy, on an individual participant level, without the spatial smoothing that is normally used in fMRI, revealing a network of brain regions active both during action observation and execution. This includes regions traditionally associated with motor control, including the posterior parietal cortex, the dorsal and ventral premotor cortex, the supplementary motor cortex, and the cerebellum. That the ventral premotor and posterior parietal cortex had this property is in line with the monkey studies showing mirror neurons in these regions. That the dorsal premotor, supplementary motor, and cerebellum had this property extended the original animal studies by providing candidate areas that could also contain mirror neurons (Keysers & Gazzola, 2009).

However, fMRI suffers from an important caveat. That a voxel is active both during action observation and action execution is compatible with the voxel containing mirror neurons, but is not a proof of the presence of mirror neurons in this voxel: a voxel contains millions of neurons and can be active in both conditions despite some of its neurons only responding during observation and others only during execution and none responding in both conditions (Gazzola & Keysers, 2009; Zaki, Wager, Singer, Keysers, & Gazzola, 2016). It thus remains essential for single cell recordings to explore the presence of mirror neurons in these regions. This has been done for the dorsal premotor cortex (Cisek & Kalaska, 2004) and the supplementary motor cortex (Mukamel, Ekstrom, Kaplan, Iacoboni, & Fried, 2010), confirming in both cases the presence of mirror neurons.

In addition to these regions involved in motor programming, voxels in the somatosensory system also show shared activations, including voxels in the posterior sector of the primary somatosensory cortex (Brodmann Area 1 and 2) and in the secondary somatosensory cortex (Gazzola & Keysers, 2009). Importantly, the pattern of activity across voxels in these regions contains specific information about what action someone else is performing, as revealed by multivoxel pattern classification (Etzel, Gazzola, & Keysers, 2008). This triggered the realization that the brain of the witness may not only trigger motor programs that would enable not only the performance of similar actions, but also somatosensory representations of what it would feel like to move in the observed way (Keysers, Kaas, & Gazzola, 2010). The activation of somatosensory cortices while witnessing the actions of others has been confirmed in a number of fMRI studies (Caspers, Zilles, Laird, & Eickhoff, 2010).

In addition, studies have also shown that sensing touch on one's own skin and witnessing someone else being touched triggers brain activity in overlapping voxels in the secondary somatosensory cortex (Keysers et al., 2004), adding evidence for the involvement of the somatosensory system in processing the somatosensory experiences of others.

Jointly, this data provides a potential neural correlate for kinesthetic empathy: when we witness the movements of others, our brain creates a representation of that movement by triggering representations of the motor programs and somatosensory feedback we would normally experience when performing similar movements or experiencing similar tactile input. The synaptic wiring necessary to map the sight of such movements to their somatosensory and motor representations could be acquired via Hebbian learning (Del Giudice, Manera, & Keysers, 2009; Keysers and Gazzola, 2014b; Keysers & Perrett, 2004; Keysers, Perrett, & Gazzola, 2014).

In a recent experiment, we further explored if these somatosensory and motor representations are functionally segregated in the brain, or strongly interact during action observation. We perturbed activity in the somatosensory cortex using transcranial magnetic stimulation (TMS) while measuring responses in the rest of the brain. We found that TMS-induced changes in brain activity in the somatosensory cortex caused a change in brain activity in the motor cortices as well (including the ventral and dorsal premotor cortex), showing that rather than generating separate motor and somatosensory representations of other people's actions, the brain seems to create an integrated somatosensory–motor representation

(Valchev, Gazzola, Avenanti, & Keysers, 2016). This is a striking match to the concept of kinesthetic empathy put forward by philosopher Theodor Lipps to explain how we enjoy art by feeling ourselves into the object of our contemplation, performing an inner mimesis, feeling ourselves moving as if we were the character depicted in the art, and thereby literally moved (Lipps, 1903).

Viewing facial expressions also triggers pattern of activity in somatosensory and motor structures that resemble the pattern when producing similar facial expressions (van der Gaag, Minderaa, & Keysers, 2007). In addition, viewing bodily expressions of emotions triggers activity in the motor system of the observer (Borgomaneri, Gazzola, & Avenanti, 2012), including activations in the premotor and parietal regions involved in moving the body (de Gelder et al., 2010). Accordingly, these motor structures could contain representations of the actions of others that could play a key role in helping us decode what other people are feeling (Jabbi & Keysers, 2008; Keysers, 2011).

THE NEURAL CORRELATES OF EMOTIONAL EMPATHY

Much of what we know about how our brain allows us to share the emotions of others takes its route from a number of fMRI studies in the early 2000s that employed an experimental design similar to those that had mapped the putative human mirror neuron system and employing a simple rationale: If empathy involves triggering a particular emotion in the witness because of witnessing that emotion in an object, a brain region involved in this ability should be activated both when the participant experiences a specific emotion himself and when witnessing the same emotion experienced by another. The first published experiment to use this logic induced the emotion of disgust in participants using unpleasant odors and then let the same participants witness the disgusted facial expressions of others. Voxels involved in experiencing and witnessing disgust were then identified separately, and the two maps were overlaid. Two brain regions had voxels that were common to witnessing and experiencing disgust (Wicker et al., 2003). The first was the anterior insula (AI) at the transition with the inferior frontal gyrus. The second was the rostral cingulate cortex (rCC) at the border between what is often referred to as the mid and anterior cingulate. The second experiment had participants experience a mild electric shock to induce pain in the self, and showed the same participants a visual cue that indicated that their romantic partner, who was also in the scanner room, received an electroshock (Singer et al., 2004). Again, the AI and rCC had voxels activated in both cases. Numerous studies have since confirmed the vicarious activation of the

AI and rCC during witnessing the emotions of others, be it for disgust (Jabbi, Bastiaansen, & Keysers, 2008; Jabbi & Keysers, 2008; Jabbi, Swart, & Keysers, 2007) or pain (Lamm, Decety, & Singer, 2011).

A key observation of that literature is that participants that report experiencing more empathy in their lives, as captured by pen-and-paper self-report questionnaires, show stronger activation of the AI and rCC when witnessing the disgust (Jabbi et al., 2007) and pain (Engen & Singer, 2013; Lamm et al., 2011; Singer et al., 2004) of others. This correlation between AI and rCC activity and empathy scores has been used as a key argument to suggest that the AI and rCC activations might be the neural basis for the empathy reported by participants. In addition, psychopaths, who are known for reduced empathy, also show reduced activation levels in these structures while perceiving the pain of others (Meffert, Gazzola, den Boer, Bartels, & Keysers, 2013). Whether the same is true for autism remains less clear (Bird et al., 2010; Hadjikhani et al., 2014).

The AI and rCC have been found to be involved in the experience and witnessing of aversive emotions such as disgust and pain, independently of how these emotions are witnessed and experienced, leading to considering them the core component of affect sharing (Engen & Singer, 2013; Lamm et al., 2011). Based on how witnesses infer the emotions of others, these regions receive their functional input through different routes. When the emotion is witnessed via facial expressions, as mentioned earlier, ventral premotor regions in the inferior frontal gyrus and somatosensory cortices that overlap with the regions involved in generating similar facial expressions are activated (van der Gaag et al., 2007), and these regions seem to provide the main functional connectivity to the AI (Jabbi & Keysers, 2008). If the states of others are inferred via linguistic narratives, these structures are embedded into a network of language-related areas that are likely to trigger the AI and rCC activation (Jabbi et al., 2008; Lerner, Honey, Silbert, & Hasson, 2011). Similarly, for pain, when pain is perceived from facial expressions or images of body parts in pain, the premotor regions of the inferior frontal gyrus are also activated (Lamm et al., 2011). In addition, if the somatic causes of pain are in the focus of the stimuli, somatosensory brain regions, including the SI and SII, are additionally activated (Keysers et al., 2010).

A much smaller number of studies have explored the neural basis of empathy for positive emotions, including joy and rewards. We have shown that experiencing and observing others experience pleasant tastes triggers activations in the gustatory cortex in the AI (Jabbi et al., 2007). We have also shown that witnessing others receiving the feedback that they had chosen a correct action overlaps with brain regions responding to similar positive outcomes for the participant's own actions, in particular in a frontostriatal network known to encore reward prediction errors, suggesting that we can learn vicariously from the actions of others by empathizing with reward prediction errors (Monfardini et al., 2013). Similarly, in a recent study, the ventral striatum was found to encode reward prediction errors independently of whether the reward was received by the self of by someone else (Lockwood, Apps, Valton, Viding, & Roiser, 2016). Hence, while the brain regions for sharing positive and negative valences differ, the brain seems equipped with systems that could endow it with the ability to share both positive and negative emotions with others.

FACTORS MODULATING AI AND rCC ACTIVATIONS IN HUMANS

Initially, empathy was studied in neuroscience as if it were a relatively automatic process, in which participants witness the emotions of others and trigger neural representations of their own emotions. This view is in line with the literature on mirror neurons in monkeys, in which activations of the monkey's own actions occur reliably within ~100 ms each time after witnessing the action of others (Keysers et al., 2003) and that had traditionally been explained in a simple, hard-wired feed-forward stream of synaptic connections (Nelissen et al., 2011). In this view, the main determinants of variance in activations are the emotions expressed in the stimuli and individual differences in empathy that would correspond to differences in the strength of the connections between sensory cortices and the AI and rCC.

More recently, a different view has emerged, which considers empathy a motivated process that participants flexibly deploy based on their self-interest (Keysers & Gazzola, 2014a,b; Zaki, 2014). In this view, in the absence of any particular motivation to empathize, participants do recruit the AI and rCC when they witness the emotions of others to a certain degree. Individuals vary in how strongly they display such spontaneous activations. However, a number of factors can dramatically down- or upregulate these spontaneous activations (Engen & Singer, 2013; Keysers & Gazzola, 2014a,b), five of which we will review here.

Fairness: the first such factor was observed by Singer et al. (2006). She had participants play a prisoner's dilemma game with two confederates prior to scanning, with one of the confederates being fair and trustworthy and the other unfair. Thereafter, she measured AI and rCC activation while participants witnessed either of them receiving shocks. She found that male participants show much lower activations when witnessing the unfair player receiving shocks, suggesting that perceived fairness can alter AI and rCC responses to witnessed pain.

In-/Outgroup: a number of studies have shown increased empathy-related activations in response to ingroup members receiving shocks compared to outgroup members. This is true if the distinction is suggested by the object having a skin color that was similar or different from that of the witness (Azevedo et al., 2013) or by the object declaring to like the same or a different football team as the witness (Hein, Silani, Preuschoff, Batson, & Singer, 2010).

Responsibility: while in most paradigms participants witness the object experience emotions that have nothing to do with the witness' own actions, more recently, paradigms have started to explore the impact of creating causal links between the witness' actions and the object's emotions. For instance Cui, Abdelgabar, Keysers, & Gazzola, 2015 created a paradigm in which both the witness and the object had to report the central character in a flanker task. The witness then saw whether he and the object performed the task correctly or not. If either of them erred, the object received an electroshock, and the witness saw the object's facial reaction via a CCTV connection. This creates situations in which the witnessed pain is (1) entirely the responsibility of the object (in cases in which the witness had performed the task correctly, but the object erred), (2) the shared responsibility of both (when both erred), or (3) entirely the witness' responsibility (when only the witness had erred). Activations in the AI and rCC were much stronger when the observed pain was entirely the responsibility of the witness compared to both other situations.

Voluntary empathy: a small number of paradigms have asked participants to deliberately up- or downregulate their empathy for the pain of others. In one of them, healthy participants and convicted psychopathic criminals first passively viewed videos of hands hurting each others and later were asked to view these video's again, but this time trying to deliberately empathize with the victims in the videos. During the initial spontaneous viewing, psychopathic individuals activated their AI, rCC, and somatosensory cortices less than the controls. When both groups were instructed to empathize, this difference disappeared (Meffert et al., 2013). This suggests that individual differences in the AI and rCC activation are not as stable as the concept of trait differences in empathy might suggest. Rather, both groups have a similar *ability* to empathize when they want to, but differ in the degree to which this ability is spontaneously deployed in a situation in which there is no incentive to empathize (Keysers & Gazzola, 2014a,b).

Prior experience: Substantial literature has shown in the domain of kinesthetic empathy that prior experience with a particular action increases activation of premotor, somatosensory, and parietal regions associated with empathizing with that action. For instance women ballet dancers show more activation for the sight of female movements, and men for the sight of male movements for which they have prior motor experience (Calvo-Merino, Grèzes, Glaser, Passingham, & Haggard, 2006). Similarly, experiencing a pressure pain prior to viewing others experience such pain alters the level of the AI and rCC activation in those witnessing that pain (Preis, Schmidt-Samoa, Dechent, & Kroener-Herwig, 2013). Again, it has been argued that Hebbian learning might be the key to associating one's own emotional states with the sight of the behavioral cues that signal such states in self and other (Del Giudice et al., 2009; Keysers & Gazzola, 2014b; Keysers & Perrett, 2004).

Together this draws a picture in which participants are endowed with a certain ability to activate the AI and rCC that can be shaped by a participant's past history of emotions, but the degree to which this ability is deployed while perceiving the emotions of others is highly flexible. From an evolutionary perspective, it is important to note that emotional contagion, empathy, and sympathy (which of these is most intimately associated with the AI and rCC activations remains unclear) are thought to be associated with costs and benefits. Empathizing with the sufferance of others by itself is aversive, and sympathy will motivate the witness to help the victim, which is typically costly. On the other hand, showing signs of empathy leads to increased liking, which may be beneficial if the target of empathy might later be in a position to help the witness. In addition, helping an ingroup member with whom one is likely to interact repeatedly may encourage reciprocity in the future. Finally, helping an offspring or kin has fitness benefits in a classic kin-selection sense. Accordingly, evolution should build mechanisms that deploy these phenomena in a motivated, context-dependent fashion to cash in on their benefits when such benefits prevail, but forfeit their costs if costs prevail. This would fit with empathy reductions for outgroup and unfair members, who are less likely to reciprocate prosocial actions, and with the ability to deploy empathy voluntarily. Finally, empathy can also play a key role in helping individuals find out which of their actions are beneficial vs. detrimental to others, which is key for individual's ability to function in groups. If empathy is seen as such a learning signal that punishes vicariously behaviors that harm others and rewards vicariously behaviors that benefit others, this signal should be strongest when the witnessed emotion is perceived as resulting of ones own actions, which could explain the modulation by responsibility and when they affect people we care about, which would explain the ingroup effects. Overall, such a functional view of empathy also sheds light on individual differences in propensity for empathy, as observed for instance when comparing psychopathic criminals with healthy volunteers, by suggesting that some individuals may be tuned toward a more exploitational lifestyle, in which empathy should not be deployed so readily, while others may be tuned to a more cooperative lifestyle, in which empathy should be deployed more readily (Keysers & Gazzola, 2014a; Meffert et al., 2013).

LIMITATION OF OUR UNDERSTANDING OF THE NEURAL BASIS OF EMPATHY FROM fMRI AND THE NEED FOR ANIMAL STUDIES

Since 2003, hundreds of studies have investigated the neural correlates of empathy in the emotional domain using fMRI, and even just the original two studies (Singer et al., 2004; Wicker et al., 2003) have attracted over 2500 citations (based on WoS in 2016). The success of these studies lies in the fact that they suggest (1) that emotional empathy may find its neural correlate in mirror-like neurons that fire when we experience an emotion and when we witness that emotion in others, and (2) that their activity (while we witness the emotions of others) confers us with the ability to feel what they feel and motivates us to help them. However, both of these core hypotheses remain untested. This is because these experiments are conducted in humans, in which it is, as we will review below, difficult to test either of them.

It cannot be systematically tested in humans that the same neurons in the AI and rCC are active during the witnessing and experiencing of emotions. A voxel if active both during the witnessing and experiencing of a particular emotion is compatible not only with the same neurons within it triggering activation in both cases, but also with different neurons in the same voxel recruited in both cases. If the latter were the case, the brain would retain a separation of the neural basis of our own emotions and those of others, which would thus fail to resemble what happens in the motor system, and thus fail to provide a convincing correlate of empathy. Although surgical procedures for the treatment of epilepsy offer occasional opportunities to record from single neurons in these regions in humans (Hutchison et al., 1999), they seldom offer the opportunity to carefully characterize the response pattern of neurons over multiple hours of testing, making it difficult to ascertain that neurons truly selectively represent a particular emotional experience (e.g., pain) during the observation and experience of an emotion. If brief testing then reveals the presence of some neurons recruited by both the experience and observation of emotions, we are left wondering what it is that these neurons make us share with others (Zaki et al., 2016). Indeed, if a neuron were to respond to pain when experienced by the self and the other, one is tempted to see it as potentially endowing us with the ability to share pain with others. However, if the same neuron responds when we experience disgust or fear, it would no longer allow us to specifically share pain with others, but rather a more general negative state of aversion. If it were to also respond to the observation and experience of a salient pleasant state (e.g., winning 100$), it would merely allow us to share arousal with others. Fine-grained characterizations of the tuning curve (i.e., to what a neuron does and does not respond) during both the observation and experience of emotions is thus key to understanding the representational mechanisms that underpin our ability to share and understand the emotions of others, but requires time scales of testing that are typically unavailable in humans.

The second assumption, namely that it is the activations of the AI and rCC while witnessing the emotions of others that allow us to share their emotions and motivates our social behavior, is equally untested because these regions are relatively deep structures, over 3 cm from the surface of the cortex. Although some rare patients exist with lesions in the AI that also show reductions in empathy and the ability to recognize disgust (Adolphs, Tranel, & Damasio, 2003; Calder, Keane, Manes, Antoun, & Young, 2000), such lesions are not restricted to a particular brain region and induce compensatory cortical plasticity that make it difficult to associate the activity of a particular region to a specific psychological function. To test the causal contribution of these regions precisely, one would need to modulate their activity focally and show that this would modulate our ability to share the emotions of others. This would be essential as well to determine whether the activation of particular brain regions might be associated specifically with particular forms of empathy—would it trigger a mere emotional contagion, empathy proper, or sympathy? This is difficult to do, because the state-of-the-art methods for noninvasive brain manipulation in humans, including TMS or transcranial direct current stimulation (tDCS), are currently unable to modulate brain activity so deep without having larger effects on the cortices that are closer to the surface.

Finally, we have observed over the last decade a multitude of individual differences in empathy-like phenomena across individuals and patient groups. What causes underlie these differences? What synaptic plasticity mechanisms may differ across individuals to account for differences in the association between the perception of their states and the emotions we feel as a result? What genes associated with these differences truly cause these differences?

Accordingly, the past decade of human neuroscience of empathy has played a key role in developing strong hypotheses about the brain regions and mechanisms that might be involved in generating our subjective sense of what other people do and feel. This work has also pointed out a number of situational factors that modulate these brain activations. Testing these hypotheses, however, now urgently requires us to develop animal models of empathy in which we can leverage the powerful techniques of animal neuroscience to image (e.g., using genetically encoded calcium indicators such as GCamP6), record (e.g., using silicon probes in freely moving animals), and modulate (e.g., using chemical and optogenetic activations and deactivations) the cellular basis of these phenomenon and generate a truly mechanistic understanding of these phenomena.

FUNDING

Christian Keysers is supported by the European Research Council of the European Commission (ERC-StG-312511) and the Netherlands Organization for Scientific Research (VICI grant 453-15-009). Valeria Gazzola is supported by the Netherlands Organization for Scientific Research (VIDI grant 452-14-015), and the Brain and Behavior Research Foundation (NARSAD young investigator grant 22453).

References

Adolphs, R., Tranel, D., & Damasio, A. R. (2003). Dissociable neural systems for recognizing emotions. *Brain and Cognition, 52*(1), 61–69. doi: 10.1016/S0278-2626(03)00009-5.

Atsak, P., Orre, M., Bakker, P., Cerliani, L., Roozendaal, B., Gazzola, V., & Keysers, C. (2011). Experience modulates vicarious freezing in rats: A model for empathy. *PLoS One, 6*(7), e21855. doi: 10.1371/journal.pone.0021855.

Azevedo, R. T., Macaluso, E., Avenanti, A., Santangelo, V., Cazzato, V., & Aglioti, S. M. (2013). Their pain is not our pain: Brain and autonomic correlates of empathic resonance with the pain of same and different race individuals. *Human Brain Mapping, 34*(12), 3168–3181. doi: 10.1002/hbm.22133.

Ben-Ami Bartal, I., Decety, J., & Mason, P. (2011). Empathy and pro-social behavior in rats. *Science (New York, NY), 334*(6061), 1427–1430. doi: 10.1126/science.1210789.

Bird, G., Silani, G., Brindley, R., White, S., Frith, U., & Singer, T. (2010). Empathic brain responses in insula are modulated by levels of alexithymia but not autism. *Brain, 133*(5), 1515–1525. doi: 10.1093/brain/awq060.

Borgomaneri, S., Gazzola, V., & Avenanti, A. (2012). Motor mapping of implied actions during perception of emotional body language. *Brain Stimulation, 5*(2), 70–76. doi: 10.1016/j.brs.2012.03.011.

Calder, A. J., Keane, J., Manes, F., Antoun, N., & Young, A. W. (2000). Impaired recognition and experience of disgust following brain injury. *Nature Neuroscience, 3*(11), 1077–1078. doi: 10.1038/80586.

Calvo-Merino, B., Grèzes, J., Glaser, D. E., Passingham, R. E., & Haggard, P. (2006). Seeing or doing? Influence of visual and motor familiarity in action observation. *Current Biology: CB, 16*(19), 1905–1910. doi: 10.1016/j.cub.2006.07.065.

Caspers, S., Zilles, K., Laird, A. R., & Eickhoff, S. B. (2010). ALE meta-analysis of action observation and imitation in the human brain. *NeuroImage, 50*(3), 1148–1167. doi: 10.1016/j.neuroimage.2009.12.112.

Cisek, P., & Kalaska, J. F. (2004). Neural correlates of mental rehearsal in dorsal premotor cortex. *Nature, 431*(7011), 993–996. doi: 10.1038/nature03005.

Cui, F., Abdelgabar, A. R., Keysers, C., & Gazzola, V. (2015). Responsibility modulates pain-matrix activation elicited by the expressions of others in pain. *NeuroImage, 114*, 371–378. doi: 10.1016/j.neuroimage.2015.03.034.

Decety, J., & Jackson, P. L. (2004). The functional architecture of human empathy. *Behavioral and Cognitive Neuroscience Reviews, 3*(2), 71–100. doi: 10.1177/1534582304267187.

Del Giudice, M., Manera, V., & Keysers, C. (2009). Programmed to learn? The ontogeny of mirror neurons. *Developmental Science, 12*(2), 350–363. doi: 10.1111/j.1467-7687.2008.00783.x.

Engen, H. G., & Singer, T. (2013). Empathy circuits. *Current Opinion in Neurobiology, 23*(2), 275–282. doi: 10.1016/j.conb.2012.11.003.

Etzel, J. A., Gazzola, V., & Keysers, C. (2008). Testing simulation theory with cross-modal multivariate classification of fMRI data. *PLoS One, 3*(11), e3690. doi: 10.1371/journal.pone.0003690.

Fogassi, L., Ferrari, P. F., Gesierich, B., Rozzi, S., Chersi, F., & Rizzolatti, G. (2005). Parietal lobe: From action organization to intention understanding. *Science (New York, NY)*, *308*(5722), 662–667. doi: 10.1126/science.1106138.

van der Gaag, C., Minderaa, R. B., & Keysers, C. (2007). Facial expressions: What the mirror neuron system can and cannot tell us. *Social Neuroscience*, *2*(3–4), 179–222. doi: 10.1080/17470910701376878.

Gallese, V., Fadiga, L., Fogassi, L., & Rizzolatti, G. (1996). Action recognition in the premotor cortex. *Brain: A Journal of Neurology*, *119*(Pt 2), 593–609. doi: 10.1093/brain/119.2.593.

Gazzola, V., & Keysers, C. (2009). The observation and execution of actions share motor and somatosensory voxels in all tested subjects: Single-subject analyses of unsmoothed fMRI data. *Cerebral Cortex*, *19*(6), 1239–1255. doi: 10.1093/cercor/bhn181.

de Gelder, B., Van den Stock, J., Meeren, H. K. M., Sinke, C. B. A., Kret, M. E., & Tamietto, M. (2010). Standing up for the body Recent progress in uncovering the networks involved in the perception of bodies and bodily expressions. *Neuroscience & Biobehavioral Reviews*, *34*(4), 513–527. doi: 10.1016/j.neubiorev.2009.10.008.

Graziano, M. S. A. (2016). Ethological action maps: A paradigm shift for the motor cortex. *Trends in Cognitive Sciences* doi: 10.1016/j.tics.2015.10.008.

Greene, J. T. (1969). Altruistic behavior in the albino rat. *Psychonomic Science*, *14*(1), 47–48. doi: 10.3758/BF03336420.

Hadjikhani, N., Zürcher, N. R., Rogier, O., Hippolyte, L., Lemonnier, E., Ruest, T., & Prkachin, K. M. (2014). Emotional contagion for pain is intact in autism spectrum disorders. *Translational Psychiatry*, *4*(November 2013), e343. doi: 10.1038/tp.2013.113.

Hein, G., Silani, G., Preuschoff, K., Batson, C. D., & Singer, T. (2010). Neural responses to ingroup and outgroup members' suffering predict individual differences in costly helping. *Neuron*, *68*(1), 149–160 10.1016/j.neuron.2010.09.003.

Hutchison, W. D., Davis, K. D., Lozano, A. M., Tasker, R. R., & Dostrovsky, J. O. (1999). Pain-related neurons in the human cingulate cortex. *Nature Neuroscience*, *2*, 403–405.

Jabbi, M., Bastiaansen, J., & Keysers, C. (2008). A Common anterior insula representation of disgust observation, experience and imagination shows divergent functional connectivity pathways. *PLoS One*, *3*(8), e2939. doi: 10.1371/journal.pone.0002939.

Jabbi, M., & Keysers, C. (2008). Inferior frontal gyrus activity triggers anterior insula response to emotional facial expressions. *Emotion (Washington, DC)*, *8*(6), 775–780. doi: 10.1037/a0014194.

Jabbi, M., Swart, M., & Keysers, C. (2007). Empathy for positive and negative emotions in the gustatory cortex. *NeuroImage*, *34*(4), 1744–1753. doi: 10.1016/j.neuroimage.2006.10.032.

Jola, C., Ehrenberg, S., & Reynolds, D. (2012). The experience of watching dance: Phenomenological–neuroscience duets. *Phenomenology and the Cognitive Sciences*, *11*(1), 17–37. doi: 10.1007/s11097-010-9191-x.

Keysers, C. (2011). *The Empathic Brain*. Amsterdam, The Netherlands: Social Brain Press.

Keysers, C., & Gazzola, V. (2007). Integrating simulation and theory of mind: From self to social cognition. *Trends in Cognitive Sciences*, *11*(5), 194–196. doi: 10.1016/j.tics.2007.02.002.

Keysers, C., & Gazzola, V. (2009). Expanding the mirror: Vicarious activity for actions, emotions, and sensations. *Current Opinion in Neurobiology*, *19*(6), 666–671. doi: 10.1016/j.conb.2009.10.006.

Keysers, C., & Gazzola, V. (2014a). Dissociating the ability and propensity for empathy. *Trends in Cognitive Sciences*, *18*(4), 163–166. doi: 10.1016/j.tics.2013.12.011.

Keysers, C., & Gazzola, V. (2014b). Hebbian learning and predictive mirror neurons for actions, sensations and emotions. *Philosophical Transactions of the Royal Society B: Biological Sciences*, *369*(1644), 20130175. doi: 10.1098/rstb.2013.0175.

Keysers, C., Kaas, J. H., & Gazzola, V. (2010). Somatosensation in social perception. *Nature Reviews. Neuroscience*, *11*(6), 417–428. doi: 10.1038/nrn2833.

Keysers, C., Kohler, E., Umiltà, M. A., Nanetti, L., Fogassi, L., & Gallese, V. (2003). Audiovisual mirror neurons and action recognition. *Experimental Brain Research*, *153*(4), 628–636. doi: 10.1007/s00221-003-1603-5.

Keysers, C., & Perrett, D. I. (2004). Demystifying social cognition: A Hebbian perspective. *Trends in Cognitive Sciences, 8*(11), 501–507. doi: 10.1016/j.tics.2004.09.005.

Keysers, C., Perrett, D. I., & Gazzola, V. (2014). Hebbian learning is about contingency, not contiguity, and explains the emergence of predictive mirror neurons. *Behavioral and Brain Sciences, 37*(2), 205–206. doi: 10.1017/S0140525X13002343.

Keysers, C., Wicker, B., Gazzola, V., Anton, J. -L., Fogassi, L., & Gallese, V. (2004). A touching sight: SII/PV activation during the observation and experience of touch. *Neuron, 42*(2), 335–346 Retrieved from http://www.ncbi.nlm.nih.gov/pubmed/15091347.

Kohler, E., Keysers, C., Umiltà, M. A., Fogassi, L., Gallese, V., & Rizzolatti, G. (2002). Hearing sounds, understanding actions: Action representation in mirror neurons. *Science (New York, NY), 297*(5582), 846–848. doi: 10.1126/science.1070311.

Lamm, C., Decety, J., & Singer, T. (2011). Meta-analytic evidence for common and distinct neural networks associated with directly experienced pain and empathy for pain. *Neuro-Image, 54*(3), 2492–2502. doi: 10.1016/j.neuroimage.2010.10.014.

Lerner, Y., Honey, C. J., Silbert, L. J., & Hasson, U. (2011). Topographic mapping of a hierarchy of temporal receptive windows using a narrated story. *The Journal of Neuroscience: The Official Journal of the Society for Neuroscience, 31*(8), 2906–2915. doi: 10.1523/JNEURO-SCI.3684-10.2011.

Lipps, T. (1903). *Asthetik: Psychologie des Schönen und der Kunst*. Hamburg: Leopold Voss.

Lockwood, P. L., Apps, M. A. J., Valton, V., Viding, E., & Roiser, J. P. (2016). Neurocomputational mechanisms of prosocial learning and links to empathy. *Proceedings of the National Academy of Sciences of the United States of America, 113*(35), 9763–9768. doi: 10.1073/pnas.1603198113.

Meffert, H., Gazzola, V., den Boer, J. A., Bartels, A. A. J., & Keysers, C. (2013). Reduced spontaneous but relatively normal deliberate vicarious representations in psychopathy. *Brain: A Journal of Neurology, 136*(Pt 8), 2550–2562. doi: 10.1093/brain/awt190.

Monfardini, E., Gazzola, V., Boussaoud, D., Brovelli, A., Keysers, C., & Wicker, B. (2013). Vicarious neural processing of outcomes during observational learning. *PLoS One, 8*(9), e73879. doi: 10.1371/journal.pone.0073879.

Mukamel, R., Ekstrom, A. D., Kaplan, J., Iacoboni, M., & Fried, I. (2010). Single-neuron responses in humans during execution and observation of actions. *Current Biology, 20*(8), 750–756. doi: 10.1016/j.cub.2010.02.045.

Nelissen, K., Borra, E., Gerbella, M., Rozzi, S., Luppino, G., Vanduffel, W., & Orban, G. A. (2011). Action observation circuits in the macaque monkey cortex. *The Journal of Neuroscience: The Official Journal of the Society for Neuroscience, 31*(10), 3743–3756. doi: 10.1523/JNEUROSCI.4803-10.2011.

Panksepp, J., & Panksepp, J. B. (2013). Toward a cross-species understanding of empathy. *Trends in Neurosciences*doi: 10.1016/j.tins.2013.04.009.

Preis, M. A., Schmidt-Samoa, C., Dechent, P., & Kroener-Herwig, B. (2013). The effects of prior pain experience on neural correlates of empathy for pain: An fMRI study. *Pain, 154*(3), 411–418. doi: 10.1016/j.pain.2012.11.014.

Rice, G. E., & Gainer, P. (1962). "Altruism" in the albino rat. *Journal of Comparative and Physiological Psychology, 55*, 123–125 Retrieved from http://www.ncbi.nlm.nihgov/pubmed/14491896.

Singer, T., Seymour, B., O'Doherty, J., Kaube, H., Dolan, R. J., & Frith, C. D. (2004). Empathy for pain involves the affective but not sensory components of pain. *Science (New York, NY), 303*(5661), 1157–1162 10.1126/science.1093535.

Singer, T., Seymour, B., O'Doherty, J. P., Stephan, K. E., Dolan, R. J., & Frith, C. D. (2006). Empathic neural responses are modulated by the perceived fairness of others. *Nature, 439*, 466–469.

Valchev, N., Gazzola, V., Avenanti, A., & Keysers, C. (2016). Primary somatosensory contribution to action observation brain activity-combining fMRI and cTBS. *Social Cognitive and Affective Neuroscience*doi: 10.1093/scan/nsw029.

de Vignemont, F., & Singer, T. (2006). The empathic brain: How, when and why? *Trends in Cognitive Sciences, 10*(10), 435–441. doi: 10.1016/j.tics.2006.08.008.

Wicker, B., Keysers, C., Plailly, J., Royet, J. P., Gallese, V., & Rizzolatti, G. (2003). Both of us disgusted in My insula: The common neural basis of seeing and feeling disgust. *Neuron*, *40*(3), 655–664 Retrieved from http://www.ncbi.nlm.nih.gov/pubmed/14642287.

Wispé, L. (1986). The distinction between sympathy and empathy: To call forth a concept, a word is needed. *Journal of Personality and Social Psychology*, *50*(2), 314–321. doi: 10.1037/0022-3514.50.2.314.

Zaki, J. (2014). Empathy: A motivated account. *Psychological Bulletin*, *140*(6), 1608–1647. doi: 10.1037/a0037679.

Zaki, J., Wager, T. D., Singer, T., Keysers, C., & Gazzola, V. (2016). The anatomy of suffering: Understanding the relationship between nociceptive and empathic pain. *Trends in Cognitive Sciences*, *20*(4), 249–259. doi: 10.1016/j.tics.2016.02.003.

5

Ethological Approaches to Empathy in Primates

Zanna Clay, Elisabetta Palagi***,
Frans B.M. de Waal[†,‡]

*Durham University, Durham, United Kingdom; **Natural History
Museum, University of Pisa, Pisa, Italy; †Emory University, Atlanta, GA,
United States; ‡Living Links, Yerkes National Primate Research Center,
Atlanta, GA, United States

INTRODUCTION

As our closest living relatives, nonhuman primates (henceforth primates) provide a critical window through which we can explore the evolutionary origins of human empathy. The comparative approach, based on the principle of homology, is especially relevant for informing us about aspects of hominid evolution that do not leave clear fossil traces. Given that cognitive and emotional processes underlying empathy do not fossilize, identifying homologies in the behaviors and capacities in our closest living relatives enables us to examine their evolutionary origins. In this way our primate relatives represent crucial living models for making reconstructions of our last common ancestor and for identifying uniquely human features.

Here, we review recent research exploring behavioral evidence for empathy in primates, focusing primarily on how studies using the ethological approach can inform us about the cognitive and emotional layers underling empathy. Since Darwin and the pioneering work of biologists such as Tinbergen, Lorenz, and von Frisch, the study of animal behavior under natural occurring conditions has provided critical insights into understanding the evolution of behavior. Principally grounded in observation, some of the most important ideas and theoretical constructs about evolution and behavior have been developed using the ethological approach. In this way,

Neuronal Correlates of Empathy. http://dx.doi.org/10.1016/B978-0-12-805397-3.00005-X

insights gained from natural observations must be taken in tandem with those from controlled experiments to gain a balanced and ecologically valid understanding of the evolutionary origins of empathy.

AN EVOLUTIONARY APPROACH TO EMPATHY

Definitions of empathy commonly emphasize two aspects: the sharing of emotions (emotional empathy) and the adoption of another's viewpoint (cognitive empathy) (Preston & de Waal, 2002). Empathy allows the organism to quickly relate to the states of others, which is essential for the regulation of social interactions, coordinated activity, and cooperation toward shared goals. Even though the cognitive capacity of perspective-taking assists in this, it is secondary to a more basic form of emotional synchrony. As Hoffman (1981, p. 79) noted: "Humans must be equipped biologically to function effectively in many social situations without undue reliance on cognitive processes."

Despite the existence of these dual components, a mostly cognitive definition, closer to Theory-of-Mind, has become more dominant. Accordingly, empathy is considered to be a way of gaining access to another's mind by stepping into their "shoes" (Goldman, 2006). Despite the role of cognition in human empathy, insights from neuroscience, genetics, development, and comparative psychology have all demonstrated that the beginnings of empathy are likely to involve much simpler mechanisms that are shared across mammals (de Waal, 2012; Decety, 2011). The directness of the empathy mechanism is reflected in the widely supported Perception Action Mechanism (PAM; Preston & de Waal, 2002) as a proximate explanation of empathy, which assumes that another individual's affective state is represented at least in part in the self's own experience of the same state.

All social animals need to coordinate movements and collectively respond to danger, communicate, and respond to others in need. Responsiveness to the behavioral states of conspecifics ranges from basic behavioral synchrony, rapid spread of emotional states (i.e., emotional contagion) to more cognitive forms of assessment, such as the sympathetic concern great apes show to socially close conspecifics in distress (de Waal, 2008, 2011). While adaptive, the more basic and reflex-like transmission of emotion is unlikely to involve much underlying cognition. Correspondingly, emotional contagion is a widely observed phenomenon occurring across many social mammals and birds (de Waal, 2011), which emerges early in human infancy (Decety, 2010; Hoffman, 2008). In contrast, a great ape (Clay & de Waal, 2013a) approaching another individual in distress to offer friendly contact is more discriminating, and involves not only a direct emotional response, but also an assessment of the reason for the distress as well as a motivation to ameliorate the situation (de Waal, 2008).

The selection pressure to evolve rapid emotional connectedness likely started in the context of parental care, which across evolutionary time, became extended to become an adaptation to group living (Preston, 2013). By signaling their emotional states, infants promote their caregiver to respond to their needs, such as feeding and offering warmth or shelter. Offspring signals are not just responded to but induce an emotional state, suggestive of parental distress at the perception of offspring distress (MacLean, 1985). This connection may explain the role of oxytocin in consolation behavior (Burkett et al., 2016) as well as higher rates of empathy responses in females than males (Zahn-Waxler, Robinson, & Emde, 1992).

During both ontogeny and evolution, learning and intelligence add layers of complexity to the basic state-matching mechanism of empathy, making the response increasingly discerning until full-blown emotional and cognitive empathy emerges (de Waal, 2012). Comparative research on animals has been pivotal in elucidating how the layering of these different capacities is organized, as well as for determining which aspects of empathy are common across species, which are found in a select number of large-brained animals and which are uniquely human.

EMPATHY IN GREAT APES AND MONKEYS: A TAXONOMIC DIVIDE?

A central question in the study of primate empathy is whether great apes represent a genuine taxonomic divide in their expressions of empathy as compared to monkeys, or whether the differences are more graded. This is because great apes, along with some other large-brained animals and birds appear to exhibit more sophisticated forms of cognitive empathy than other animals. Great apes appear to be able to put themselves in someone else's shoes, such as by offering comfort to victims of aggression to alleviate their distress, a capacity known as consolation (chimpanzees: de Waal & van Roosmalen, 1979; Romero, Castellanos, & de Waal, 2010; bonobos: Clay & de Waal, 2013a,b; Palagi, Paoli, & Borgognini Tarli, 2004, gorillas: Cordoni, Palagi, & Tarli, 2006; ravens: Fraser & Bugnyar, 2010; canids: Cools, Van Hout, & Nelissen, 2008; elephants: Plotnik & de Waal, 2014). Consolation is often contrasted with other forms of post-conflict affiliation provided by bystanders in monkey and other animals, which appear to be driven by different underlying mechanisms that fulfill more directly self-serving interests (Aureli & de Waal, 2000).

In the past three decades, the advanced perspective-taking capacities of great apes have been the focus of a huge amount of experimental research, primarily to elucidate what great apes know about what others know and how they might use this knowledge (Call & Tomasello, 2008; Premack & Woodruff, 1978; Tomasello, 2014). Cumulatively, the scientific literature suggests that although great apes do not possess a full blown Theory of

Mind, they can nevertheless reason about the actions of others, perceive them as intentional beings and to some extent, attribute mental states (Call & Tomasello, 2008; Hare, Call, & Tomasello, 2001).

While it is likely that some qualitative taxonomic differences exist in the empathic and perspective-taking abilities of great apes as compared to monkeys (de Waal & Aureli, 1996), more recent data from monkeys, which we discuss in a following section, suggest more similarities than previously assumed (Palagi et al., 2014a).

EXPERIMENTAL AND NATURALISTIC APPROACHES TO PRIMATE EMPATHY

Here, we overview research that explores evidence for empathy in primates, both the more basic emotional aspects as well as the more cognitively sophisticated components.

Contagious Yawning

Yawn contagion is a basic form of empathic responding shared by humans, primates, and many other animals (Norscia & Palagi, 2011). Due to its evolutionary antiquity, the yawn appears morphologically identical across many different vertebrate taxa meaning that it can be easily detected and quantified across species (Baenninger, 1997; Provine, 2005). Most importantly, its plesiomorphic nature provides a common platform from which it is possible to quantitatively evaluate the capacity of animals to emotionally synchronize with others (Guggisberg, Mathis, Schnider, & Hess, 2010).

According to neurobiological (Haker, Kawohl, Herwig, & Rössler, 2013), psychological (Platek et al., 2005), and ethological findings (Campbell & de Waal, 2011; Norscia & Palagi, 2011; Palagi, Leone, Mancini, & Ferrari, 2009; Romero et al., 2010; Tan et al., 2017) yawn contagion reflects the most basal layer of empathy, which is reflexive bodily synchronization (e.g., see the "Russian Doll Model" in de Waal, 2008; Preston & de Waal, 2002). Yawn contagion can be reliably observed in naturalistic scenarios, which permits the collection of systematic data that is freed from artificial biases potentially induced in experimental settings (Bartholomew & Cirulli, 2014; Provine, 2005).

One of the most relevant findings from the ethological approach is that yawn contagion is strongly predicted by social closeness. This is consistent with research showing that human and animal empathy are biased toward individuals who are more similar, familiar, or socially bonded (Preston & de Waal, 2002). The presence of this empathic gradient distribution of yawn contagion has so far been demonstrated in humans

(Norscia & Palagi, 2011), chimpanzees (Campbell & de Waal, 2011), bonobos (Demuru & Palagi, 2012), and gelada baboons, (*Theropithecus gelada*) (Palagi et al., 2009). To test some hypotheses on the potential species differences of yawn contagion, Palagi et al. (2014b) directly compared yawn contagion in humans and bonobos and found that in both species, yawn contagion was highest between strongly bonded subjects whereas between species, yawn contagion was higher in humans than in bonobos when the two subjects were kin or friends. Nevertheless, other studies have demonstrated that bonobo's yawn contagion may be less constrained by social closeness than other species. For example, while chimpanzees show a strong in-group bias in yawn contagion (Campbell & de Waal, 2011); one recent study showed that bonobos showed yawn contagion in response to both familiar and stranger conspecifics, highlighting their apparently xenophilic nature (Tan et al., 2017). Overall, the cross-species approach supports that the multilayered architecture of empathy is shared across many species (de Waal, 2008; Preston & de Waal, 2002). Monkeys, such as gelada baboons also demonstrate yawn contagion with those that they shown high levels of social closeness (Palagi et al., 2009). In short, yawn contagion represents an instance of truly affective reactions that may be mediated by neural pathways of old evolutionary origin and provides a cornerstone for motor mimicry. Taken together, the results suggest that the relationship between yawn contagion and empathy may have developed earlier than the last common ancestor between monkeys and apes, including humans.

Rapid Facial Mimicry

In humans, the sharing of emotions through rapid and unconscious facial mimicry (RFM) is another important mechanism for emotional and empathic contagion (Dimberg, Thunberg, & Elmehed, 2000). Considering the importance that RFM plays in regulating human social interactions, it was proposed that this phenomenon was more widespread than previously thought. In nonhuman animals, RFM was first demonstrated in orangutans (*Pongopygmaeus*) (Davila-Ross, Menzler, & Zimmermann, 2008) and later in gelada baboons (Mancini, Ferrari, & Palagi, 2013). As with yawn contagion, RFM has also been shown to promote dyadic social bonding by facilitating emotional connection and social affiliation. In geladas, as in humans, RFM involving mother–infant dyads was more evident than between unrelated subjects (Mancini et al., 2013). Such affective matching is at the basis of infants' neuropsychological maturation and maternal attachment.

The larger facial display repertoires of tolerant species (Dobson, 2012) and their cooperative tendencies during play favor the expression of the phenomenon of RFM, which is linked to emotional contagion. In agreement with this hypothesis, Scopa & Palagi (2016) demonstrated that *Macaca fuscata* (Japanese macaque, a despotic species) and *M. tonkeana* (Tonkean macaque, a highly tolerant species) strongly differ in the performance of this behavior. Although both species performed play faces at comparable levels, only Tonkean macaques showed RFM, and also played for longer periods. Taking together, the results suggest that RFM promotes communicative exchanges and the behavioral coordination of play sequences.

Recent research with great apes using eye-tracking and physiological techniques further asserts the role that basic behavioral and physiological mechanisms play in modulating the emotional components of empathy. Recently, Kano, Hirata, Deschner, Behringer & Call (2016) used noninvasive thermo-imaging techniques to explore chimpanzee's emotional responsiveness to playbacks of conspecific agonistic aggression. Following presentation of the playbacks, the chimpanzees experienced a pronounced drop in nasal temperature, indicative of an emotional response. Although still in its infancy, thermo-imaging presents a promising technique with which the emotional responses of animals can be systematically explored. Likewise, recent advances in eye-tracking research has been fruitful in elucidating species differences in emotional responding. Two convergent studies, for example, have demonstrated that bonobos, as compared to chimpanzees, are more attuned to emotional information, spending more time attending to emotional stimuli (Kret et al., 2016), and showing more interest in the faces and eyes of conspecifics (Kano, Hirata, & Call, 2015). Such differences correspond to behavioral (Hare, Melis, Woods, Hastings, & Wrangham, 2007; Herrmann, Hare, Call, & Tomasello, 2010) and neuroanatomical differences (Rilling et al., 2012; Stimpson et al., 2015) suggesting that bonobos are more attuned to social and emotional information, as well as being more socially tolerant and less aggressive than chimpanzees.

Sympathetic Concern

Being able to experience another individual's emotions, separate from one's own, is considered a more cognitively demanding form of empathy, known as sympathetic concern (Preston & de Waal, 2002; Zahn-Waxler et al., 1992). In primates, the spontaneous offering of comfort behaviors to victims in distress, that is, consolation (de Waal & van Roosmalen, 1979), is widely accepted as a marker of empathy. Its status as an empathic behavior stems from the early studies of human consolation behavior by Zahn-Waxler et al. (1992) as well as evidence that this behavior improves the psychological and emotional state of the victims by reducing their anxiety without necessarily providing any direct benefits to the actor (Clay & de Waal, 2013a; Romero et al., 2010). Consolation requires both the emotional perception of anothers' affective state as well as spontaneous response of offering a friendly gesture toward the victim in distress. As described in earlier sections, consolation appears to be rare across the animal kingdom, having been demonstrated in great apes (de Waal & van Roosmalen, 2013; Romero et al., 2010; bonobos: Palagi et al., 2004; Clay & de Waal, 2013a,b; gorillas: Cordoni et al., 2006), as well as corvids (Fraser & Bugnyar, 2010), canids (Cools et al., 2008), elephants (Plotnik & de Waal, 2014), and more recently in voles (Burkett et al., 2016).

Consistent with an empathy-based explanation, consolation is predicted by social closeness, with individuals more likely to console closely bonded individuals and kin (Clay & de Waal, 2013a; Palagi & Norscia, 2013; Romero et al., 2010). Its calming effect has also been demonstrated in a number of studies, generally through observations of reduced rates of self-scratching, a valid behavioral indicator of anxiety in human and primates (Clay & de Waal, 2013a; Fraser, Stahl, & Aureli, 2008; Palagi & Norscia, 2013; Romero et al., 2010).

Whereas most studies of animal consolation have been observational, on which little neuroscience can be conducted, the underlying neural mechanisms are important as they may provide an answer as to how this behavior comes about. Is consolation based on a calculation about social consequences or reciprocity, as some have suggested (Fraser, Koski, Wittig, & Aureli, 2009), or is it more closely tied to caring responses and attachment, as the empathy hypothesis would suggest? A recent study on voles by Burkett et al. (2016) finally allowed the social neuroscience that could answer this question. Burkett and co-workers observed consolation between pair-bonded male and female voles following a stressful separation. They found several critical pieces of evidence in favor of the empathy hypothesis, such as: (1) the social bias also reported for other empathy expressions (the voles directed this behavior to mates and siblings but not strangers), (2) emotional contagion (a nearly perfect match between the physiological stress levels of the male and female involved even though only one of the two was experimentally stressed), and (3) the role of oxytocin (i.e., the behavior was eliminated by blocking oxytocin receptors in the brain). The study concluded that consolation behavior rests on caring and attachment neural circuitry that is shared by all mammals.

Consolation is distinct from affiliative contact sought out by the victim in that the bystander actively offers reassurance after a conflict in which they played no role. Recent research with captive bonobos investigated the outcome of spontaneous versus solicited postconflict affiliative contacts (Palagi & Norscia, 2013), which showed that while both forms of affiliation provided protection to victims against further attacks only spontaneous comfort improved their emotional state, as demonstrated by reduced levels of self-scratching following consolation but not victim-initiated contact. Hence, the improvement of the emotional state of the victim seems to depend on the spontaneity of being offered the contact rather than the contact itself.

Comparative research with primates has also demonstrated that empathic responding, such as consolation, relates to the level of tolerance characterized in different primate societies (Palagi et al., 2014a,b). By directly comparing across macaque species, Palagi and colleagues found that *Macaca tonkeana* (highly tolerant) but not *Macaca fuscata* (highly despotic) performed consolation. Given the close phylogenetic relationship, it is likely that both species have similar empathic capacities, but only the more tolerant species express it via consolation. Maternal style is probably at the basis of the difference in tolerance levels shown by the various species of macaques. The permissiveness of mothers is crucial to determine the tolerance propensity in adults. The social inhibition characterizing despotic species makes the mothers more protective toward their infants thus reducing their opportunity to engage in social contacts. Thierry (1985) demonstrated that infants' social canalization is not present in more tolerant species such as Tonkean macaques, whose mothers allow their infants to freely interact with other group members. These early ontogenetic differences are enduring and persist later in the dynamics of adult–adult social interactions, thus probably shaping the different social styles observed in macaque species.

Targeted Helping

As with consolation, another important spontaneous manifestation of more advanced empathic perspective-taking is so-called targeted helping, which is help fine-tuned to another's specific situation (de Waal, 1996). This kind of helping requires that one individual understands the predicament another one is in and provides the exact solution to the other's problem. Targeted helping requires a shift in perspective, because an individual needs to move beyond being sensitive to others toward an explicit other-orientation.

For descriptions of spontaneous cases of targeted helping, mostly among apes, dolphins, and elephants, see de Waal (2010). Yamamoto, Humle, & Tanaka (2012) recently addressed targeted helping in a controlled experiment with chimpanzees. A chimpanzee learned two alternative ways to obtain juice, which involved either using a rake or using a straw. At test, while no tools were directly available, a chimpanzee in an adjacent testing room was provided with both tools, among a set of other tools. In a direct display of targeted helping, the chimpanzee placed in the adjacent cage, provided the chimpanzee with the correct tool they needed for the task by handing it through a window. Importantly, the observer chimpanzee only provided the correct tool if they had directly seen the other's situation; otherwise, tools were picked at random. This experiment demonstrated not only that chimpanzees are willing to assist each other, but that they could also take the other's specific needs into account.

SPONTANEOUS RESPONSES TO NATURALLY OCCURRING EMOTIONAL EVENTS

Rare and spontaneous emotional events, such as births, injuries, and deaths, provide a valuable opportunity to explore how animals both respond to and perceive the emotional needs of others.

The sociality and emotional involvement around birth has been proposed as a unique feature of our species and traces its origin to the very beginning of human evolution. Although humans differ from other primates for the imperative need for assistance during the delivery, other primates nevertheless also show a certain degree of sociality around birth.

For instance, Demuru et al. (2015) systematically observed three daytime births in bonobos, which occurred within the social group. Analyses highlighted the extraordinary social event that birth represents to bonobos, especially in regards to the females. Females stayed closer to the parturient than males, followed and inspected her, and engaged in affiliative interactions. The more dominant and elderly females also provided a sort of "assistance" to the parturient, by performing a manual gesture to capture the infant as it was being born. A birth event in this species seems to acquire a special social value that reinforces female social relationships, which are at the basis of bonobo sociality (Furuichi, 2011). As a whole, these data suggest that in complex societies birth is a collective emotional event whose social dynamics mirror those typical of the species. The female collective support in bonobos might be one of the building blocks of more complex and culturally shaped forms of sociality, which are expressed during the delivery event in humans.

Responses to injuries or toward group mates unable to free themselves from human-laid traps also provide relevant insights into underlying empathic processes, with some cases providing fairly convincing evidence of "targeted helping." Primates typically show strong interest in the injuries of other group members, and frequently tend to the victim by licking the wound and inspecting it using soft touch. Recently, Tokuyama et al. (2012) reported a case of members of a wild bonobo community searching for and attempting to assist a lost group member injured by a snare. After becoming caught by the snare, other group members attempted to unfasten the snare from his fingers, as well as gently inspecting and attending to his wound. Interestingly, after leaving the injured individual after having been unable to free him, the group returned the following day to this precise location and made visible efforts to search for him. While anecdotal, such examples of targeted helping in response to others in need are indicative of empathy. Interestingly, such cases are not confined to great apes, with a recent report of wild black-fronted titi monkeys showing evidence of empathic responding to an injured conspecific, who was even an outgroup member (Clyvia, Kaizer, Santos, Young, & Cäsar, 2014).

EMPATHY DEVELOPMENT AND EMOTION REGULATION

Research on human development has shown that consolation and other empathy-based behaviors increase with age, both in terms of the frequency and type of targets (Light & Zahn-Waxler, 2011). For example, children first comfort family members and then other children, especially when hurt; with two years being considered a key developmental milestone for empathy and prosocial development (Decety, 2011; Zahn-Waxler et al., 1992). Nevertheless, while more complex forms of cognitive empathy emerge in conjunction with other cognitive skills, the foundations for other-orientated empathetic responding appear to be already in place from an early age. For example, although it was previously thought that infants show primarily self-oriented and reflexive emotional responses toward others (i.e., exhibiting personal distress in response to someone else's distress), research with 8–16-month infants has shown that responses of personal distress toward other's distress were actually rare (Roth-Hanania, Davidov, & Zahn-Waxler, 2011). In this way, even at an early age, human infants appear to be equipped with the emotional architecture that will then develop to become increasingly more attuned to responding to the emotional states of others.

While research into emotion development in primates is still in its infancy, there is increasing evidence of homologous patterns. Recent data on one-year old children (Davidov, Zahn-Waxler, Roth-Hanania, & Knafo, 2013), young bonobos (Clay & de Waal, 2013a,b) and more recently young chimpanzees (Webb et al., 2017) indicate that the empathic sphere can emerge and expand earlier than previously thought. In the study of bonobos, consolation was more likely to be offered by younger bystanders, especially juveniles compared to adults, suggesting that mechanisms for empathic responding are already in place in this great apes from an early age (Clay & de Waal, 2013a,b).

Importantly, the study described earlier also demonstrated the important role that early attachment experiences play in the development of empathy in human and primates. Studies with human infants consistently show that disruptions in development, brought on by neglect/deprivation or abuse, negatively affect the development of empathic behavior, attachment, and emotion regulation (De Bellis, 2005). In the study of bonobos, which took place at a sanctuary that rescues and rehabilitates orphans from the bush-meat trade, mother-reared juveniles were more likely to offer consolation than age-matched orphaned juveniles. Consistent with patterns observed in human infants, these findings highlight the role of rearing and early attachment in emotional development and empathy. Since the seminal work of Harlow, it has been shown (van Ijzendoorn, Bard, Bakerman, Kranenburg, & Ivan, 2009) that maternal care in infancy is critical for the development of secure and organized attachment styles as well as for cognitive and socioemotional development. Congruent with human studies, which indicate that empathy and emotion regulation are negatively impacted by early trauma and disruptions in development (De Bellis, 2005), the authors found that juvenile bonobos displaying higher levels of socioemotional competence, emotion regulation, and sociability were also more likely to offer empathic concern to individuals in distress (Clay & de Waal, 2013b).

To investigate under which social and environmental conditions consolation develops along the ontogenetic trajectory of humans, Cordoni and Palagi (in prep) carried out an ethological study on preschool children using the same observational methodology used for primates. The study revealed many similarities in patterns of consolation observed in great apes and human children, which included the timing of consolation as well as the fact that children offered comfort to others at a young age. As in the great apes, consolation seems to be a spontaneous and immediate response toward distress in others, and was effective in reducing victim anxiety, as has been shown in primates (Clay & de Waal, 2013a; Romero et al., 2010). In conclusion, when data on humans and apes are collected using similar procedures and operational definitions striking similarities in the sympathetic response can emerge.

CONCLUSIONS

In summary, through a combination of observational and experimental methods, the ethological approach provides many important insights into the evolutionary origins and development of empathy. Based on the principle of homology, the comparative approach provides a crucial tool for understanding and unraveling the complexities of empathy, particularly for elucidating the emergence of its various cognitive and affective layers. Overall, the emerging picture is that the affective components of empathy are already deeply rooted in our primate past, as demonstrated for example, by evidence of emotional synchrony in behaviors such as yawning and RFM. In terms of cognitive empathy, we see some taxonomic shifts in the more cognitively sophisticated forms of empathy observed in great apes and other large-brained animals, namely sympathetic concern and targeted helping. Importantly, evolutionary shifts toward more human-like empathy are congruent with the ontogenetic shifts in human development. This notion of continuity is further asserted by evidence of early forms of sympathetic concern present in both young human infants and some monkey species. Future research will need to determine the extent to which the underlying mechanisms driving other-oriented responding in human infants are similar to that observed in monkeys and/or great apes.

References

Aureli, F., & Waal, F. B. M. (2000). *Natural conflict resolution*. University of California Press.

Baenninger, R. (1997). On yawning and its functions. *Psychonomic Bulletin & Review*, 4(2), 198–207.

Bartholomew, A. J., & Cirulli, E. T. (2014). Individual variation in contagious yawning susceptibility is highly stable and largely unexplained by empathy or other known factors. *PLoS One*, 9(3), e91773.

Burkett, J., Andari, E., Johnson, Z., Curry, D., de Waal, F. B. M., & Young, L. (2016). Oxytocin-dependent consolation behavior in rodents. *Science*, 351(6271), 375–378.

Call, J., & Tomasello, M. (2008). Does the chimpanzee have a theory of mind? 30 years later. *Trends in Cognitive Sciences*, 12(5), 187–192.

Campbell, M. W., & de Waal, F. B. M. (2011). Ingroup-outgroup bias in contagious yawning by chimpanzees supports link to empathy. *PLoS One*, 6(4), e18283.

Clay, Z., & de Waal, F. B. M. (2013a). Bonobos respond to distress in others: Consolation across the age spectrum. *PLoS One*, 8(1), e55206.

Clay, Z., & de Waal, F. B. M. (2013b). Development of socio-emotional competence in bonobos. *Proceedings of the National Academy of Sciences U.S.A.*, 110(45), 18121–18126.

Clyvia, A., Kaizer, M., Santos, R., Young, R., & Cäsar, C. (2014). Do wild titi monkeys show empathy? *Primate Biology*, 1(1), 23.

Cools, A. K., Van Hout, A. J. M., & Nelissen, M. H. (2008). Canine reconciliation and third-party-initiated postconflict affiliation: Do peacemaking social mechanisms in dogs rival those of higher primates? *Ethology*, 114(1), 53–63.

Cordoni, G., Palagi, E., & Tarli, S. B. (2006). Reconciliation and consolation in captive western gorillas. *International Journal of Primatology*, 27(5), 1365–1382.

Davidov, M., Zahn-Waxler, C., Roth-Hanania, R., & Knafo, A. (2013). Concern for others in the first year of life: Theory, evidence, and avenues for research. *Child Development Perspectives*, 7(2), 126–131.

Davila-Ross, M., Menzler, S., & Zimmermann, E. (2008). Rapid facial mimicry in orangutan play. *Biology Letters*, 4(1), 27–30.

de Bellis, M. D. (2005). The psychobiology of neglect. *Child Maltreatment*, 10(2), 150–172.

de Waal, F. B. M (1996). *Good natured*. Harvard University Press.

de Waal, F. B. M. (2008). Putting the altruism back into altruism: the evolution of empathy. *Annual Reviews of Psychology*, 59, 279–300.

de Waal, F. B. M (2010). *The age of empathy: Nature's lessons for a kinder society*. Broadway Books.

de Waal, F. B. M. (2011). Empathy in primates and other mammals. *Empathy: From bench to bedside*. 87.

de Waal, F. B. M. (2012). The antiquity of empathy. *Science*, 336(6083), 874–876.

de Waal, F. B. M., & Aureli, F. (1996). Consolation, reconciliation, and a possible cognitive difference between macaques and chimpanzees. *Reaching into thought: The minds of the great apes*. 80-110.

de Waal, F. B. M., & van Roosmalen, A. (1979). Reconciliation and consolation among chimpanzees. *Behavioural Ecology and Sociobiology*, 5, 55–66.

Decety, J. (2010). The neurodevelopment of empathy in humans. *Developmental Neuroscience*, 32(4), 257–267.

Decety, J. (2011). The neuroevolution of empathy. *Annals of the New York Academy of Sciences*, 1231(1), 35–45.

Demuru, E., & Palagi, E. (2012). In bonobos yawn contagion is higher among kin and friends. *PLoS One*, 7(11), e49613.

Demuru, E., Ferrari, P. F., & Palagi, E. (2015). Birth in bonobos (*Pan paniscus*): female cohesiveness and emotional sharing. *Folia Primatologica*, 86, 271.

Dimberg, U., Thunberg, M., & Elmehed, K. (2000). Unconscious facial reactions to emotional facial expressions. *Psychological Science*, 11(1), 86–89.

Dobson, S. D. (2012). Coevolution of facial expression and social tolerance in macaques. *American Journal of Primatology, 74*(3), 229–235.

Fraser, O. N., & Bugnyar, T. (2010). Do ravens show consolation? Responses to distressed others. *PLoS One, 5*(5), e10605.

Fraser, O. N., Koski, S. E., Wittig, R. M., & Aureli, F. (2009). Why are bystanders friendly to recipients of aggression? *Communicative & Integrative Biology, 2*(3), 285–291.

Fraser, O. N., Stahl, D., & Aureli, F. (2008). Stress reduction through consolation in chimpanzees. *Proceedings of the National Academy of Sciences, 105*(25), 8557–8562.

Furuichi, T. (2011). Female contributions to the peaceful nature of bonobo society. *Evolutionary Anthropology: Issues, News, and Reviews, 20*(4), 131–142.

Goldman, A. I. (2006). *Simulating minds: The philosophy, psychology, and neuroscience of mindreading.* Oxford University Press.

Guggisberg, A. G., Mathis, J., Schnider, A., & Hess, C. W. (2010). Why do we yawn? *Neuroscience & Biobehavioral Reviews, 34*(8), 1267–1276.

Haker, H., Kawohl, W., Herwig, U., & Rössler, W. (2013). Mirror neuron activity during contagious yawning—an fMRI study. *Brain Imaging and Behavior, 7*, 28–34.

Hare, B., Call, J., & Tomasello, M. (2001). Do chimpanzees know what conspecifics know? *Animal Behaviour, 61*(1), 139–151.

Hare, B., Melis, A. P., Woods, V., Hastings, S., & Wrangham, R. (2007). Tolerance allows bonobos to outperform chimpanzees on a cooperative task. *Current Biology, 17*(7), 619–623.

Herrmann, E., Hare, B., Call, J., & Tomasello, M. (2010). Differences in the cognitive skills of bonobos and chimpanzees. *PLoS One, 5*(8), e12438.

Hoffman, M. L. (1981). Is altruism part of human nature? *Journal of Personality and Social Psychology, 40*(1), 121.

Hoffman, M. L. (2008). Empathy and prosocial behavior. *Handbook of Emotions, 3*, 440–455.

Kano, F., Hirata, S., & Call, J. (2015). Social attention in the two species of Pan: Bonobos make more eye contact than chimpanzees. *PLoS One, 10*(6), e0129684.

Kano, F., Hirata, S., Deschner, T., Behringer, V., & Call, J. (2016). Nasal temperature drop in response to a playback of conspecific fights in chimpanzees: A thermo-imaging study. *Physiology & Behaviour, 155*, 83–94.

Kret, M. E., Jaasma, L., Bionda, T., & Wijnen, J. G. (2016). Bonobos (*Pan paniscus*) show an attentional bias toward conspecifics' emotions. *Proceedings of the National Academy of Sciences U.S.A, 113*(14), 3761–3766.

Light, S., & Zahn-Waxler, C. (2011). 7 nature and forms of empathy in the first years of life. *Empathy: From bench to bedside*, 109.

MacLean, P. D. (1985). Brain evolution relating to family, play, and the separation call. *Archives of General Psychiatry, 42*(4), 405–417.

Mancini, G., Ferrari, P. F., & Palagi, E. (2013). Rapid facial mimicry in geladas. *Scientific Reports, 3*, 1527.

Norscia, I., & Palagi, E. (2011). Yawn contagion and empathy in *Homo sapiens. PLoS One, 6*(12), e28472.

Palagi, E., Dall'Olio, S., Demuru, E., & Stanyon, R. (2014a). Exploring the evolutionary foundations of empathy: consolation in monkeys. *Evolution and Human Behaviour, 35*(4), 341–349.

Palagi, E., Leone, A., Mancini, G., & Ferrari, P. (2009). Contagious yawning in gelada baboons as a possible expression of empathy. *Proceedings of the National Academy of Sciences U.S.A., 106*(46), 19262–19267.

Palagi, E., & Norscia, I. (2013). Bonobos protect and console friends and kin. *PLoS One, 8*(11), e79290.

Palagi, E., Norscia, I., & Demuru, E. (2014b). Yawn contagion in humans and bonobos: Emotional affinity matters more than species. *PeerJ, 2*, e519.

Palagi, E., Paoli, T., & Borgognini Tarli, S. (2004). Reconciliation and consolation in captive bonobos (*Pan paniscus*). *American Journal of Primatology, 62*(1), 15–30.

Platek, S. M., Mohamed, F. B., & Gallup, G. G. (2005). Contagious yawning and the brain. *Cognitive Brain Research, 23*(2), 448–452.

Plotnik, J. M., & de Waal, F. B. (2014). Asian elephants (*Elephas maximus*) reassure others in distress. *PeerJ, 2*, e278.

Premack, D., & Woodruff, G. (1978). *Does the chimpanzee have a theory of mind? Behavioural and Brain Sciences, 1*(04), 515–526.

Preston, S. D. (2013). The origins of altruism in offspring care. *Psychological Bulletin, 139*(6), 1305.

Preston, S. D., & de Waal, F. (2002). Empathy: Its ultimate and proximate bases. *Behavioural and Brain Sciences, 25*(01), 1–20.

Provine, R. R. (2005). Yawning: The yawn is primal, unstoppable and contagious, revealing the evolutionary and neural basis of empathy and unconscious behavior. *American Scientist, 93*(6), 532–539.

Rilling, J. K., Scholz, J., Preuss, T. M., Glasser, M. F., Errangi, B. K., & Behrens, T. E. (2012). Differences between chimpanzees and bonobos in neural systems supporting social cognition. *Social Cognitive and Affective Neuroscience, 7*(4), 369–379.

Romero, T., Castellanos, M. A., & de Waal, F. B. (2010). Consolation as possible expression of sympathetic concern among chimpanzees. *Proceedings of the National Academy of Sciences U. S. A., 107*(27), 12110–12115.

Roth-Hanania, R., Davidov, M., & Zahn-Waxler, C. (2011). Empathy development from 8 to 16 months: early signs of concern for others. *Infant Behavior and Development, 34*(3), 447–458.

Scopa, C., & Palagi, E. (2016). Mimic me while playing! Social tolerance and rapid facial mimicry in macaques (*Macaca tonkeana* and *Macaca fuscata*). *Journal of Comparative Psychology, 130*(2), 153.

Stimpson, C. D., Barger, N., Taglialatela, J. P., Gendron-Fitzpatrick, A., Hof, P. R., Hopkins, W. D., et al. (2015). Differential serotonergic innervation of the amygdala in bonobos and chimpanzees. *Social Cognitive and Affective Neuroscience.* nsv128.

Tan, J., Ariely, D., & Hare, B. (2017). Bonobos respond prosocially toward members of other groups. *Scientific Reports, 7*(1), 14733.

Thierry, B. (1985). Social development in three species of macaque (*Macaca mulatta, M. fascicularis M. tonkeana*): a preliminary report on the first ten weeks of life. *Behavioural Processes, 11*(1), 89–95.

Tokuyama, N., Emikey, B., Bafike, B., Isolumbo, B., Iyokango, B., Mulavwa, M. N., et al. (2012). Bonobos apparently search for a lost member injured by a snare. *Primates, 53*(3), 215–219.

Tomasello, M. (2014). *A natural history of human thinking.* Harvard University Press.

van Ijzendoorn, M. H., Bard, K. A., Bakerman, S., Kranenburg, M. J., & Ivan, K. (2009). Enhancement of attachment and cognitive development of young nursery-reared chimpanzees in responsive versus standard care. *Developmental Psychobiology, 51*, 173–185.

Webb, C. E., Romero, T., Franks, B., & de Waal, F. B. (2017). Long-term consistency in chimpanzee consolation behaviour reflects empathetic personalities. *Nature Communications, 8*(1), 292.

Yamamoto, S., Humle, T., & Tanaka, M. (2012). Chimpanzees' flexible targeted helping based on an understanding of conspecifics' goals. *Proceedings of the National Academy of Sciences U. S. A., 109*(9), 3588–3592.

Zahn-Waxler, C., Robinson, J. L., & Emde, R. N. (1992). The development of empathy in twins. *Developmental Psychology, 28*(6), 1038.

6

Mirror Neurons, Embodied Emotions, and Empathy

Pier F. Ferrari, Gino Coudé

Institut des Sciences Cognitives Marc Jeannerod, CNRS, Cedex, France

INTRODUCTION

Living in social environment requires the capacity to be in tune with other individuals. To thrive in a social world, animals as well as humans must have the ability to understand what other group members are doing, to read their emotions, and infer their intentions. Social animals have evolved both the capacity to display and the ability to read signals about physiological or emotional states, as well as decode others' intentions. The expression of emotion was one of Darwin's most important contributions to explain the universal principles of the theory of evolution (Darwin, 1872). Facial expressions convey important information regarding internal states and the associated physiological changes.

Several scholars agree that at the basis of empathic responses, there is an emotional response that is shared between two or more individuals, named emotional contagion (de Vignemont & Singer, 2006; Preston & de Waal, 2002). This phenomenon is probably based on an action-perception mechanism and is widespread among primates.

Empathy thus can be defined as the ability to understand and share the internal states of others. Several scholars agree that it is a complex, multidimensional phenomenon that includes a number of functional processes, including emotion recognition, emotion contagion, and emotion priming (for reviews, see Decety & Jackson, 2006; Jackson, Meltzoff, & Decety, 2006; Singer, 2006; Walter, 2011), as well as the abilities to react to the internal states of others, and to distinguish between one's own and others' internal states (Tomova, Von Dawans, Heinrichs, Silani, & Lamm, 2014). Empathy can take various forms along a spectrum. At one end of this spectrum, mimicry and emotional contagion appear to be shared by several

Neuronal Correlates of Empathy. http://dx.doi.org/10.1016/B978-0-12-805397-3.00006-1

mammalian species such as primates, mice, pigs, and dogs (Tramacere & Ferrari, 2016). At the other end of this spectrum, higher forms of empathy such as cognitive empathy rely on a conscious, deliberative process through which inferences can be made about others' bodily and affective states, beliefs, and intentions—often referred to as"mentalizing"—(Keysers & Fadiga, 2008; Zaki & Ochsner, 2012). Several studies demonstrate that humans are capable of inhibiting internal states and emotional responses that reflect those of others. Several contextual features, such as social distance, trustworthiness, group membership, and attention, can modulate empathic responses. Less clear is whether other social species, such as apes, may possess such cognitive forms of empathy. To summarize, empathy involves an affective component and a cognitive component.

These two forms of empathy are also associated with different brain networks and neurobiological mechanisms: affective empathy is associated with activity in premotor-parietal, temporal, and subcortical regions classically associated with movement, sensation, and emotion, while neural systems involved in cognitive control and decision-making, such as the cingulate, prefrontal, and temporal areas are often activated during tasks requiring cognitive empathy (Zaki & Ochsner, 2012). In the current treatise we will mainly focus on the affective components of empathy as they have been widely investigated in numerous species and because they may offer a more comprehensive view of the basic mechanisms involved.

EMPATHY, EMOTIONS, AND IMITATION

One approach to understanding the mechanism and processes involved in affective empathy is to investigate the emotional responses of animals and humans and their physiological underpinnings. Studies on emotions in animals and humans have produced a bulk of evidence showing that emotional states are often transmitted unwittingly to observers. This transmission involves prereflective processes, even though humans seem capable of consciously and unconsciously modulating them. For instance, the observation of someone laughing could be contagious. Yawning is contagious too. Several species of primates have been reported to yawn after seeing a conspecific yawn (Palagi, Norscia, & Demuru, 2014).This phenomenon is known as "emotional contagion." According to some scholars (de Waal, 2008; Palagi, Leone, Mancini, & Ferrari, 2009), the transmission of emotional states is a basic form of empathy. The fact that empathy exists in the animal world has stimulated the field to study the possible neural substrate by which empathic behavior can be engaged. Several studies in animals and humans have showed that observing another individual experiencing a sensation involves the same neural structures as when that sensation is directly experienced, a phenomenon termed "shared representation" (Lamm, Decety, & Singer, 2011; Lockwood, 2016). Such an internal simulation process is supported by action-related brain networks that activate similar sensorimotor programs during the observation and the execution of an action. This matching has been revealed in studies on pain and on empathy for pain. For instance it has been shown in humans through fMRI experiments that both empathy for pain and experiencing pain first-hand recruits the anterior cingulate cortex and the anterior insula, two brain regions known to be involved in pain perception and in the autonomic visceromotor responses associated with that specific experience (Gu, Hof, Friston, & Fan, 2013; Lockwood, 2016).

From a neurobiological perspective, an action–perception matching mechanism has been proposed to mediate some basic forms of affective empathy, where neurons resonate with the motor and the affective states of other individuals with biological similarity. Many scholars believe that the mirror neurons, or at least a mirroring mechanism, can account for some basic forms of affective empathy. In this review, we are describing how the self-other matching, subserved by mirror neurons, is at the origin of a basic form of emotion sharing.

UNDERSTANDING THE ACTIONS OF OTHERS FROM "INSIDE"

The discovery that the perception of other individuals' behavior recruits neurons involved in the execution of the same action, have had important implications for social neuroscience research. Mirror neurons were first described in the ventral premotor cortex of the monkey (di Pellegrino, Fadiga, Fogassi, Gallese, & Rizzolatti, 1992) and subsequently in the convexity of the inferior parietal lobule (Fogassi et al., 2005). They have the key characteristic of firing both when a specific action is executed and when the same action, performed by another individual, is observed (di Pellegrino et al., 1992; Gallese, Fadiga, Fogassi, & Rizzolatti, 1996; Rizzolatti & Craighero, 2004; Rizzolatti, Fadiga, Gallese, & Fogassi, 1996). The fact that mirror neurons are found in motor cortical areas has led to the idea that others' actions can be translated into a motor code exploiting one's own action knowledge (Iacoboni, Molnar-Szakacs, Gallese, Buccino, & Mazziotta, 2005; Keysers & Fadiga, 2008). Such translation would allow an observer to understand the action through an implicit mapping of others' actions onto his or her own motor representation of that action.

Mirror neurons are evolutionarily well preserved and have been found thus far in three different species of macaques, in marmosets—a new world monkey—and in songbirds (Ferrari & Rizzolatti, 2014). Neurons with mirror properties were also found in humans, where invasive single cell recording can be rarely used beyond the context of surgical procedures. Despite these limitations, some findings show that a mirror mechanism exists at the single-unit level in human. In one study, mirror neurons were recorded in the supplementary motor area, hippocampus, parahippocampal gyrus, and enthorinal cortex while the patients, in addition to executing and observing grasping actions, also performed facial expressions and observed the same facial expressions (Mukamel, Ekstrom, Kaplan, Iacoboni, & Fried, 2010). In another study, a few neurons in anterior cingulate cortex responded both when the patient received painful stimulation and when the patient watched the surgeon apply a painful stimulation to himself (Hutchison, Davis, Lozano, Tasker, & Dostrovsky, 1999). The human mirror neuron system is more widely studied using neuroimaging techniques. Meta-analyses of fMRI and PET experiments reveal that regions of the human brain that are homologues of those monkey mirror system show increased activity in overlapping areas during both execution and observation of actions (Caspers, Zilles, Laird, & Eickhoff, 2010; Molenberghs, Cunnington, & Mattingley, 2012). Although the activation patterns found—in the inferior frontal gyrus, inferior parietal lobule, and ventral premotor cortex—cannot be directly attributed to the activation of mirror neurons, similar activation patterns during action–execution and action–observation support the idea of a neural mirroring mechanism and of the existence of a mirror system in human.

In itself, the existence of mirror neurons, or of a mirroring mechanism, has indeed important implications for social neuroscience and empathy research. For example, activity of some mirror neurons is modulated by the goal of an action, the intentions of another individual, or by social cues such as gaze direction, which is critical for social understanding (Bonini et al., 2011; Fogassi et al., 2005). Moreover, some mirror neurons are multimodal, that is, they fire while seeing an action, hearing the sound of an action, or both (Kohler et al., 2002). The complex responses of these neurons are specific to the type of action. For instance, they responded to a peanut breaking when the action was only observed, only heard, or both heard and observed, and did not respond to the vision and sound of another action. This multimodal matching is a feature that can account for the fact that during empathic experiences, subjects can activate shared motor representations by exploiting multiple sensory channels, including visual, tactile, and auditory, thus making the shared experience with another individual much richer and more complex. By the fact that mirror neurons account for such a wide range of parameters, one can infer that mirror neurons provide access to a rich input, and that this input might be useful for the observer's capability to understand others' actions and share their experience.

FACE MIRRORING AND SHARING EMOTION

The description of a mirror mechanism has not been limited to actions that, like those involving the hand, do not require any significant involvement of affective processes. In particular, in recent decades, studies on the mirror mechanism explored another fundamental dimension of social behavior and cognition: emotions. Living in social groups requires not only the ability to gather information about the mental or emotional states of others, but also to share aspects of others' internal states and subjective feelings and emotions. In such contexts, a class of mirror neurons are of particular significance for sharing emotion: mouth mirror neurons (Ferrari, Gallese, Rizzolatti, & Fogassi, 2003). Mouth mirror neurons were found in the ventral premotor cortex of macaques and are of two types, either driven by ingestive actions, or more interestingly, triggered by communicative actions performed with the mouth like lipsmacking (a typical macaque gesture related to affiliative behavior). These communicative mirror neurons were the first single-unit data recorded from the classical mirror neuron system (ventral premotor cortex and inferior parietal lobule) suggesting that the postulated mechanism by which a mapping of the observed action onto an internal motor representation could be extended to the emotional domain. In human, neuroimaging studies have shown that areas of the MN system are activated during the observation and imitation of facial expressions (Carr, Iacoboni, Dubeau, Mazziotta, & Lenzi, 2003; Montgomery & Haxby, 2008; van der Gaag, Minderaa, & Keysers, 2007). In addition to the mirror neuron areas, the insular cortex and the anterior cingulate cortex were found to be activated during the observation of emotional facial expressions (Carr et al., 2003; Hennenlotter et al., 2005; van der Gaag et al., 2007; Wicker et al., 2003) suggesting that other brain structures that are associated with emotion processing also have mirroring properties. Interestingly, these brain areas are involved in integrating internal-visceral signals and to activate autonomic responses as well as to decode the emotional valence of a stimuli, either social or nonsocial (Eisenberger, 2015).

The face conveys key information about the physiological and emotional state of an individual and allows an observer to access the emotional status of that individual. Both human and nonhuman primates use facial expressions to communicate their emotions and intentions. In a social context, facial expressions are produced either to initiate an exchange, or to respond to others' emotional signal. Thus, the orofacial communication channel is one by which the basic manifestations of empathy are commonly expressed in animals and humans.

From an evolutionary perspective, orofacial communicative behavior probably has its origin in the mother–infant dyad. Mammalian maternal care necessitates that the mother detect hunger or discomfort in their young, and this has probably resulted in an evolutionary pressure path to develop a facial communication channel useful to get attuned with one another. During face-to-face communication, mothers of several primate species have been described to mimic their infants, and exaggerate gestures and vocalization to express positive affects, solicit infants positive engagement and regulate infants' emotions (Dettmer et al., 2016; Ferrari et al., 2009; Stern, 1985; Trevarthen & Aitken, 2001). Despite their immature brains and limited cognitive skills, infants demonstrate active interest in their social world. They show a surprising ability to discriminate adult communicative expressions and to imitate. The capacity to mimic others' behaviors and emotions seems to stem from an ancient evolutionary capacity that is already present very early in primate development. For example, human, ape, and monkey neonates are capable of imitating facial gestures displayed by a human model (Bard, 2007; Ferrari et al., 2006; Meltzoff & Moore, 1977). This capacity probably evolved to tune an infant's behavior to that of the mother, thus facilitating the mother–infant relationship and imitative exchanges (Ferrari et al., 2006; Paukner, Simpson, Ferrari, Mrozek, & Suomi, 2014).

These early imitative responses displayed by the neonates seem to reflect the activity of a mirror mechanism (Ferrari et al., 2012). The imitation of tongue protrusion or lipsmacking by the macaque neonate is, in fact, associated with variations in the alpha frequency of the EEG (Ferrari et al., 2012), where the suppression of this rhythm—known as the mu rhythm—during action execution and observation has been interpreted as the result of activation of sensorimotor cortex, an indirect marker of mirror neuron activity (Fox et al., 2016). More specifically, when adults and children view others' goal-directed actions, electroencephalography activities recorded over the motor cortex are suppressed (Marshall & Meltzoff, 2011). These data show that sensorimotor structures are activated at an early age during facial gesture observation in infant monkeys (Ferrari et al., 2012), providing an important clue regarding the presence of a mirror mechanism at birth. Human infant data also show that the observation and execution of facial expressions produces significant mu event-related desynchronization (ERD) for centrally located electrodes (Rayson, Bonaiuto, Ferrari, & Murray, 2016). Emotional processing mirror mechanism has been identified in the insula and anterior cingulate cortex of the macaque (Gothard, Battaglia, Erickson, Spitler, & Amaral, 2007; Livneh, Resnik, Shohat, & Paz, 2012).

MIMICRY AND EMOTIONAL CONTAGION

Facial expression mirroring is of utmost significance for empathy research and offers a glimpse of the evolutionary path from which the basic form of empathy might have emerged. The observation that mirror mechanisms and neonatal imitation can be identified during the early postnatal period in humans and macaques, suggests that these species have evolved a neural system for emotional communicative exchanges that is functional very early in life (Ferrari et al., 2003; Mancini, Ferrari, & Palagi, 2013).

However, other forms of imitation can emerge during development, evolve from the early forms of imitation described earlier, and be maintained throughout adulthood, especially if the contexts in which they are displayed are emotionally engaging and positive.

Facial mimicry is a stimulus-driven response that aligns the motor behavior of the observer and the demonstrator (de Vignemont & Singer, 2006; Preston & de Waal, 2002; Zaki & Ochsner, 2012). This fast response appears to be the basis of emotional contagion, in which emotions spread from individual to individual through mimicry, for instance, when someone smiles and observers immediately do the same (Lakin & Chartrand, 2003).

It has been proposed that facial mimicry is a common behavioral manifestation of empathy (Davila Ross, Menzler, & Zimmermann, 2008; Niedenthal, Mermillod, Maringer, & Hess, 2010) along with contagious yawning (Campbell, Carter, Proctor, Eisenberg, & de Waal, 2009; Palagi et al., 2009), and both occur in human and nonhuman animals. Mimicry is more common among empathic people or among individual with strong bonds (Chartrand & Bargh, 1999; Mancini et al., 2013), stressing that there is a link between empathy and mimicry.

This unconscious and automatic phenomenon likely relies on brain mechanisms that facilitate the activation of shared motor representations, which may promote the emergence of a sense of familiarity and emotional connectedness between individuals (Palagi et al., 2009).

ADVANTAGES OF MIRROR NEURONS

The work reviewed here has clearly demonstrated that there is a basic mechanism of emotion mirroring in animals and humans and that it is likely responsible for basic forms of affective empathy. Mirror neurons can make us empathize with others in two possible ways. In one case, mirror neurons can simulate the observed action and, by activating other emotional brain centers, are capable of triggering activity in those brain centers and evoke the corresponding subjective feeling. In the other case, activity of mirror neurons alone can suffice to evoke a similar emotional experience and feeling. With this latter account we hypothesize that mirror properties are not limited to parietal-premotor regions where the simulation mechanism could describe the observed emotion in a motor format, but without the rich physiological/body responses that accompanies emotions. Neuroimaging studies and the neurophysiological work by Mukamel, Ekstrom, Kaplan, Iacoboni, and Fried (2010) on patients described mirror responses in areas beyond the traditional motor cortices, thus supporting the notion that perceiving others' emotions can actually recruit the same circuit involved in decoding the same emotion from a first person perspective. This concept echoes with Theodor Lipps' theoretical account of empathy or Einfühlung—literally meaning "feeling into"—in which the capacity to empathize relies on a mechanism of projecting the self into the other (Lipps, 1903). The psychological process of projection proposed by Lipps was based on imitating the inner part of the emotion ("innere Nachahmung"—inner imitation).

Why is a mirror mechanism a parsimonious explanation of the capacity of humans and other animals to share emotions? This action–perception mechanism presents several advantages when compared to other models: (1) It is a simple mechanism automatically exploiting the internal motor knowledge of the individual to recognize others' behavior; (2) It can mediate an immediate response in case of danger; (3) It can explain several phenomena of imitation and of motor social facilitation such as emotional contagion; and (4) It is an effortless mechanism that engages in the absence of more complex cognitive skills, and is therefore suitable for organisms at an early stage of development, thus supporting several behavioral processes.

References

Bard, K. A. (2007). Neonatal imitation in chimpanzees (*Pan troglodytes*) tested with two paradigms. *Animal Cognition, 10*(2), 233–242 http://doi.org/10.1007/s10071-006-0062-3.

Bonini, L., Serventi, F. U., Simone, L., Rozzi, S., Ferrari, P. F., & Fogassi, L. (2011). Grasping neurons of monkey parietal and premotor cortices encode action goals at distinct levels of abstraction during complex action sequences. *The Journal of Neuroscience: The Official Journal of the Society for Neuroscience, 31*(15), 5876–5886 http://doi.org/10.1523/JNEUROSCI. 5186-10.2011.

Campbell, M. W., Carter, J. D., Proctor, D., Eisenberg, M. L., & de Waal, F. B. M. (2009). Computer animations stimulate contagious yawning in chimpanzees. *Proceedings of the Royal Society B: Biological Sciences, 276*(1676), 4255–4259.

Carr, L., Iacoboni, M., Dubeau, M. C., Mazziotta, J. C., & Lenzi, G. L. (2003). Neural mechanisms of empathy in humans: a relay from neural systems for imitation to limbic areas. *Proceedings of the National Academy of Sciences of the United States of America, 100*(9), 5497–5502 http://doi.org/10.1073/pnas.0935845100.

Caspers, S., Zilles, K., Laird, A. R., & Eickhoff, S. B. (2010). ALE meta-analysis of action observation and imitation in the human brain. *NeuroImage, 50*(3), 1148–1167 http://doi.org/10.1016/j.neuroimage.2009.12.112.

Chartrand, T. L., & Bargh, J. a. (1999). The chameleon effect: the perception-behavior link and social interaction. *Journal of Personality and Social Psychology, 76*(6), 893–910 Available from http://www.ncbi.nlm.nih.gov/pubmed/10402679.

Darwin, C. (1872). *The expression of the emotions in man and animals.* London: John Murray.

Davila Ross, M., Menzler, S., & Zimmermann, E. (2008). Rapid facial mimicry in orangutan play. *Biology Letters, 4*(1), 27–30.

de Vignemont, F., & Singer, T. (2006). The empathic brain: how, when and why? *Trends in Cognitive Sciences, 10*(10), 435–441.

de Waal, F. B. M. (2008). Putting the altruism back into altruism: The evolution of empathy. *Annual Review of Psychology, 59*, 279–300.

Decety, J., & Jackson, P. L. (2006). A social-neuroscience perspective on empathy. *Current Directions in Psychological Science, 15*(2), 54–58.

Dettmer, A. M., Kaburu, S. S. K., Simpson, E. A., Paukner, A., Sclafani, V., Byers, K. L., et al. (2016). Neonatal face-to-face interactions promote later social behaviour in infant rhesus monkeys. *Nature Communications, 7*, 1–6.

di Pellegrino, G., Fadiga, L., Fogassi, L., Gallese, V., & Rizzolatti, G. (1992). Understanding motor events: A neurophysiological study. *Experimental Brain Research, 91*(1), 176–180.

Eisenberger, N. I. (2015). Social pain and the brain: Controversies, questions, and where to go from here. *Annual Review of Psychology, 66*(1), 601–629.

Ferrari, P. F., Gallese, V., Rizzolatti, G., & Fogassi, L. (2003). Mirror neurons responding to the observation of ingestive and communicative mouth actions in the monkey ventral premotor cortex. *European Journal of Neuroscience, 17*(8), 1703–1714 http://doi.org/10.1046/j.1460-9568.2003.02601.x.

Ferrari, P. F., Paukner, A., Ruggiero, A., Darcey, L., Unbehagen, S., & Suomi, S. J. (2009). Interindividual differences in neonatal imitation and the development of action chains in rhesus macaques. *Child Development, 80*(4), 1057–1068 http://doi.org/10.1111/j.1467-8624.2009.01316.x.

Ferrari, P. F., & Rizzolatti, G. (2014). Mirror neuron research: The past and the future. *Philosophical Transactions of the Royal Society of London. Series B, Biological Sciences, 369*(1644), 1–4.

Ferrari, P. F., Vanderwert, R. E., Paukner, A., Bower, S., Suomi, S. J., & Fox, N. a. (2012). Distinct EEG amplitude suppression to facial gestures as evidence for a mirror mechanism in newborn monkeys. *Journal of Cognitive Neuroscience, 24*(5), 1165–1172.

Ferrari, P. F., Visalberghi, E., Paukner, A., Fogassi, L., Ruggiero, A., & Suomi, S. (2006). Neonatal imitation in rhesus macaques. *PLoS Biology, 4*(9), e302.

Fogassi, L., Ferrari, P. F., Gesierich, B., Rozzi, S., Chersi, F., & Rizzolatti, G. (2005). Parietal lobe: From action organization to intention understanding. *Science, 308*(5722), 662–667.

Fox, N. A., Bakermans-Kranenburg, M. J., Yoo, K. H., Bowman, L. C., Cannon, E. N., Vanderwert, R. E., et al. (2016). Assessing human mirror activity with EEG Mu rhythm: A meta-analysis. *Psychological Bulletin, 142*(3), 291–313 Available from http://dx.doi.org/10.1037/bul0000031.supp.

Gallese, V., Fadiga, L., Fogassi, L., & Rizzolatti, G. (1996). Action recognition in the premotor cortex. *Brain: A Journal of Neurology, 119*(2), 593–609.

Gothard, K. M., Battaglia, F. P., Erickson, C. A., Spitler, K. M., & Amaral, D. G. (2007). Neural responses to facial expression and face identity in the monkey amygdala. *Journal of Neurophysiology, 97*, 1671–1683.

Gu, X., Hof, P. R., Friston, K. J., & Fan, J. (2013). Anterior insular cortex and emotional awareness. *Journal of Comparative Neurology, 521*(15), 3371–3388.

Hennenlotter, A., Schroeder, U., Erhard, P., Castrop, F., Haslinger, B., Stoecker, D., et al. (2005). A common neural basis for receptive and expressive communication of pleasant facial affect. *NeuroImage, 26*(2), 581–591.

Hutchison, W. D., Davis, K. D., Lozano, A. M., Tasker, R. R., & Dostrovsky, J. O. (1999). Pain-related neurons in the human cingulate cortex. *Nature Neuroscience, 2*(5), 403–405.

Iacoboni, M., Molnar-Szakacs, I., Gallese, V., Buccino, G., & Mazziotta, J. C. (2005). Grasping the intentions of others with one's own mirror neuron system. *PLoS Biology, 3*(3), 0529–0535.

Jackson, P. L., Meltzoff, A. N., & Decety, J. (2006). Neural circuits involved in imitation and perspective-taking. *Neuroimage, 31*(1), 429–439.

Keysers, C., & Fadiga, L. (2008). The mirror neuron system: New frontiers. *Social Neuroscience*, *3*(3-4), 193–198.

Kohler, E., Keysers, C., Umiltà, M. A., Fogassi, L., Gallese, V., & Rizzolatti, G. (2002). Hearing sounds, understanding actions: action representation in mirror neurons. *Science*, *297*(5582), 846–848.

Lakin, J. L., & Chartrand, T. L. (2003). Using nonconscious behavioral mimicry to create affiliation and rapport. *Psychological Science*, *14*(4), 334–339.

Lamm, C., Decety, J., & Singer, T. (2011). Meta-analytic evidence for common and distinct neural networks associated with directly experienced pain and empathy for pain. *Neuroimage*, *54*(3), 2492–2502.

Lipps, T. (1903). Einfühlung Innere Nachahmung und Organempfindung. *Archiv Für Gesamte Psychologie*, *1*, 465–519 (Translated as "Empathy, Inner Imitation and Sense-Feelings," in A Modern Book of Esthetics, 374–382. New York: Holt, Rinehart and Winston, 1979).

Livneh, U., Resnik, J., Shohat, Y., & Paz, R. (2012). Self-monitoring of social facial expressions in the primate amygdala and cingulate cortex. *Proceedings of the National Academy of Sciences of the United States of America*, *109*(46), 18956–18961.

Lockwood, P. L. (2016). The anatomy of empathy: Vicarious experience and disorders of social cognition. *Behavioural Brain Research*, *311*, 255–266.

Mancini, G., Ferrari, P. F., & Palagi, E. (2013). In play we trust: Rapid facial mimicry predicts the duration of playful interactions in geladas. *PLoS ONE*, *8*(6), 2–6.

Marshall, P. J., & Meltzoff, A. N. (2011). Neural mirroring systems: Exploring the EEG mu rhythm in human infancy. *Developmental Cognitive Neuroscience*, *1*(2), 110–123.

Meltzoff, A. N., & Moore, M. K. (1977). Imitation of facial and manual gestures by human neonates. *Science*, *198*(4312), 75–78 Available from https://www.cs.swarthmore.edu/~meeden/DevelopmentalRobotics/77Meltzoff_Moore_Science.pdf\nhttp://www.sciencemag.org/content/198/4312/75\nhttp://www.ncbi.nlm.nih.gov/pubmed/17741897\nhttp://www.sciencemag.org/content/198/4312/75.abstract?sid=be6cc828-bc68-4.

Molenberghs, P., Cunnington, R., & Mattingley, J. B. (2012). Brain regions with mirror properties: A meta-analysis of 125 human fMRI studies. *Neuroscience & Biobehavioral Reviews*, *36*(1), 341–349.

Montgomery, K. J., & Haxby, J. V. (2008). Mirror neuron system differentially activated by facial expressions and social hand gestures: A functional magnetic resonance imaging study. *Journal of Cognitive Neuroscience*, *20*(10), 1866–1877.

Mukamel, R., Ekstrom, A. D., Kaplan, J., Iacoboni, M., & Fried, I. (2010). Single-neuron responses in humans during execution and observation of actions. *Current Biology: CB*, *20*(8), 750–756.

Niedenthal, P. M., Mermillod, M., Maringer, M., & Hess, U. (2010). The Simulation of Smiles (SIMS) model: Embodied simulation and the meaning of facial expression. *The Behavioral and Brain Sciences*, *33*(6), 417–433.

Palagi, E., Leone, A., Mancini, G., & Ferrari, P. F. (2009). Contagious yawning in gelada baboons as a possible expression of empathy. *Proceedings of the National Academy of Sciences of the United States of America*, *106*(46), 1–6.

Palagi, E., Norscia, I., & Demuru, E. (2014). Yawn contagion in humans and bonobos: emotional affinity matters more than species. *PeerJ*, *2*, e519 http://doi.org/10.7717/peerj.519.

Paukner, A., Simpson, E. A., Ferrari, P. F., Mrozek, T., & Suomi, S. J. (2014). Neonatal imitation predicts how infants engage with faces. *Developmental Science*, *17*(6), 833–840.

Preston, S. D., & de Waal, F. B. M. (2002). Empathy: its ultimate and proximate bases. *Behavioral and Brain Sciences*, *25*, 1–72.

Rayson, H., Bonaiuto, J. J., Ferrari, P. F., & Murray, L. (2016). Mu desynchronization during observation and execution of facial expressions in 30-month-old children. *Developmental Cognitive Neuroscience*, *19*, 279–287.

Rizzolatti, G., & Craighero, L. (2004). The mirror-neuron system. *Annual Review of Neuroscience*, *27*, 169–192.

Rizzolatti, G., Fadiga, L., Gallese, V., & Fogassi, L. (1996). Premotor cortex and the recognition of motor actions. *Cognitive Brain Research, 3*(2), 131–141.

Singer, T. (2006). The neuronal basis and ontogeny of empathy and mind reading: Review of literature and implications for future research. *Neuroscience and Biobehavioral Reviews, 30*(6), 855–863.

Stern, D. N. (1985). *The interpersonal world of the Infant.* New York: Basic Books.

Tomova, L., Von Dawans, B., Heinrichs, M., Silani, G., & Lamm, C. (2014). Is stress affecting our ability to tune into others? Evidence for gender differences in the effects of stress on self-other distinction. *Psychoneuroendocrinology, 43,* 95–104.

Tramacere, A., & Ferrari, P. F. (2016). In *Faces in the mirror, from the neuroscience of mimicry to the emergence of mentalizing* (pp. 1–14). (94).

Trevarthen, C., & Aitken, K. J. (2001). Infant intersubjectivity: research, theory, and clinical applications. *Journal of Child Psychology and Psychiatry, and Allied Disciplines, 42*(1), 3–48.

van der Gaag, C., Minderaa, R. B., & Keysers, C. (2007). Facial expressions: What the mirror neuron system can and cannot tell us. *Social Neuroscience, 2*(3-4), 179–222.

Walter, H. (2011). Social cognitive neuroscience of empathy: Concepts, circuits and genes. *Neuroscience, 1,* 9–17 in press.

Wicker, B., Keysers, C., Plailly, J., Royet, J. P., Gallese, V., & Rizzolatti, G. (2003). Both of us disgusted in my insula. *Neuron, 40*(3), 655–664.

Zaki, J., & Ochsner, K. N. (2012). The neuroscience of empathy: Progress, pitfalls and promise. *Nature Neuroscience, 15*(5), 675–680.

The Neurobiological Influence of Stress in the Vole Pair Bond

Adam S. Smith*, Zuoxin Wang**
*Pharmacy School, University of Kansas, Lawrence, KS, United States;
**Florida State University, Tallahassee, FL, United States

INTRODUCTION

Social bonds are an integral facet of sociality. Attachments that mature between a mother and her child or between two romantic partners can have a profound effect on the health and well-being of each individual. These bonds can provide a source of comfort and buffer the neuroendocrine responses to stress. This can reduce the prevalence of mood disorders, improving mental health. In contrast, duress can suffocate the maturation of social bonds, in terms of the impact stress has on both neurobiological substrates and behaviors ascribed to bonding. Moreover, separation from or loss of a spouse or social partner is one of the more grievous events in life. Together, this has produced an emphasis on exploring the impact that stress has on social bonding.

Neurobiological studies have highlighted the role of various monoamines and neuropeptides in regulating stress and bond-dependent behaviors. Some effort has been made to study the neural correlates of mother-infant attachments and stable romantic relationships in humans, identifying the involvement of oxytocin (Oxt), vasopressin (Avp), and dopamine. However, much of our knowledge regarding the neurobiology of social bonds comes from the study of specific social relationships in animals. Numerous mammalian and avian models have been generated to outline several neural substrates that regulate infant-caregiver, mother-infant, and, when appropriate, father-infant attachments (Neumann, 2009). In accordance to attachment theory proposed by Bowlby (1958), these

Neuronal Correlates of Empathy. http://dx.doi.org/10.1016/B978-0-12-805397-3.00007-3

forms of social bonding require neural circuitry that regulates social recognition, motivation and reward, and affiliation and emotional processing pathways. Parent-infant attachments blunt the stress responsivity and emotionality in caregivers and promote resiliency to stressful life events and shape adult social behaviors in offspring. However, to learn about the beneficial outcomes of stable social bonds between romantic or intimate partners (known as pair bonds), focal species are more restrictive, as less than 3% of mammals are socially monogamous and establish pair bonds.

From over 2 decades of research, the socially monogamous prairie vole (*Microtus ochrogaster*) has emerged as an excellent model species for examining the neurobiology of complex social behaviors, including pair bonding. Here, we focus on the pair bond of prairie voles as a source of social support as well as the negative consequences from loss of the partner or associated bond. This chapter delves into the neurobiology of stress influencing pair bonding and highlights the vole as a model for this research.

THE PRAIRIE VOLE PAIR BOND MODEL

Prairie vole pair bonding behaviors have been extensively examined in both field research and the laboratory. The prairie vole is a socially monogamous rodent species that lives primarily in the grasslands of the central United States. Sexually naive prairie voles are highly gregarious and socially tolerant toward conspecifics. After establishing breeding pair, males and females reserve affiliative behaviors and proximity to their partner and become highly territorial, confronting and fighting other conspecifics. Field studies using radiotelemetry combined with repeated trapping offered evidence that male and female establish pairs, nesting and traveling together in the wild during both breeding and nonbreeding seasons. Bonded adult males and females remain together until one partner dies, and even then, the survivor rarely forms a new pair bond (Pizzuto & Getz, 1998). Furthermore, breeding pairs in the laboratory display behavioral hallmarks of social monogamy, including preference for their social partner over unfamiliar conspecifics, aggression toward intruding conspecifics, remaining together during gestation and biparental care of offspring, distress and social-seeking behavior during periods of separation or social loss, and stress alleviation among reunion and consoling behaviors. These behaviors propagate the formation and maintenance of pair bonding in prairie voles. Behavioral paradigms have been designed to quantify these behaviors and explore the underlying neurobiological substrates in both male and female prairie voles (Table 7.1).

TABLE 7.1   Sex Differences in the Neurobiology of Social Behavior

Sociality	Male	Female	References
Partner preference	↑HPA	↓HPA	DeVries (2002)
	⇑Avp	↑Avp, DA	Aragona and Wang (2009), Cho et al. (1999)
	↑Oxt, DA	⇑Oxt	Aragona and Wang (2009), Cho et al. (1999)
Intruder aggression	↑Avp, DA	Observed[a]	Gobrogge and Wang (2016)
Social buffering	Observed[a]	↑Oxt	Smith and Wang (2014)
Consoling behavior	↑Oxt	Observed[a]	Burkett et al. (2016)

HPA, Hypothalamic-pituitary-adrenal axis; Avp, vasopressin; Oxt, oxytocin; DA, dopamine.
[a] *No data for neurobiology.*

PARTNER PREFERENCE FORMATION

Pair bond formation is facilitated by a selective preference for a social partner, known as a partner preference, in which prairie voles seek out contact with their social partner over other conspecifics. This behavior is studied using a three-chamber partner preference test initially adopted into vole research by Dr. Sue Carter (Williams, Catania, & Carter, 1992). The partner preference test is a 3-h test in which a subject is allowed to freely roam throughout the three chambers of the apparatus which differ in the social stimuli that are present-namely, an empty cage, which acts as a nonsocial control environment, and two adjoining cages that house either the subject's social partner or an unfamiliar, opposite-sex conspecific. After 24 h of cohabitation and mating, male and female prairie voles will reliably display a partner preference, providing a time course for the initial aspects of pair bond formation.

The first studies to evaluate neurochemical substrates involved in the development of a partner preference came from research investigating the impact of social and sexual experience-prerequisites for naturally induced partner preference formation-on two closely related neuropeptidergic systems, namely, Oxt and Avp. It usually requires 18-24 h of mating for the induction of partner preference formation in prairie voles; however, extended cohabitation without mating has been shown to facilitate the same behavior in females (Williams et al., 1992). Cohabitation and mating with a female increased Avp gene expression in the bed nucleus of the stria terminalis (BNST) and decreased the density of Avp immunoreactive fibers in the lateral septum (LS), a region innervated by

Avp neurons originating from the BNST (Wang, Young, De Vries, & Insel, 1998). Furthermore, Avp administration directly into the LS induces male partner preference in the absence of mating, whereas blockade of Avp action in this region prevents mating-induced preference formation (Lim & Young, 2004). The Avp fibers found in dense networks in the LS extend ventrally into the ventral pallidum (VP), and use of an adeno-associated virus to overexpress a subtype of Avp receptor (Avp receptor 1a, Avpr1a) in the VP also facilitates partner preference in males. Given the high sequence homology among Avp and Oxt receptor subtypes and high degree of chemical similarity between both peptides, Avp and Oxt are not only ligands for their specific receptors but also cross talk with the receptors of the other peptide. In the LS, this results in partner preference formation being influenced by Avp action not only on Avp receptors but also on Oxt receptors in males. In fact, site-specific injection of an Oxt receptor antagonist in the LS will prevent preference formation. In females, exposure to male chemosensory cues alters Oxt receptor density in the accessory olfactory bulb and releases Oxt in the nucleus accumbens (NAcc) from the hypothalamic paraventricular (PVN) and supraoptic (SON) nuclei (Ross & Young, 2009; Witt, Carter, & Insel, 1991). Pharmacology and viral vector studies show that local Oxt action in the NAcc and prelimbic cortex modulates partner preference formation in females (Ross & Young, 2009). Notably, Oxt injected directly in the NAcc induces partner preferences, whereas an Oxt receptor antagonist injected in either the NAcc or prelimbic cortex can inhibit mating-induced female partner preferences. Overexpression of Oxt receptor in the NAcc of sexually naive females via viral vector-mediated gene transfer accelerates partner preference formation. Together, this describes two distinct neural circuits, one in males and the other in females, which feature Avp and Oxt as neuromodulators of partner preference formation. It should be noted that these circuits might not be exclusive to each sex. For example, male prairie voles that express a higher Oxt receptor density in the NAcc are more likely to be a paired resident as opposed to a single "wanderer" (Ophir, Gessel, Zheng, & Phelps, 2012). Further, Avp and Oxt contribute to the formation of partner preference in both sexes, but the effective doses of each neuropeptide differ between males and females (Cho, DeVries, Williams, & Carter, 1999). Interestingly, recent studies have shown that epigenetic events, such as histone acetylation, are involved in the regulation of pair bonding behavior via Oxt receptors in the NAcc in both male and female voles (Wang, Duclot, Liu, Wang, & Kabbaj, 2013).

Data from a series of pharmacological and neuroanatomical experiments have indicated that mesolimbic dopamine activity, particularly in the NAcc, is another circuit through which a partner preference is formed (Aragona & Wang, 2009). Briefly, the ventral tegmental area is a dopamine-rich brain region that innervates the NAcc and creates synapses with neurons that express two families of dopamine receptors, D1-like (DRD1) and D2-like (DRD2) receptors. This neural circuit regulates the motivational value of environmental stimuli, generating adaptive goal-directed behaviors. Mating releases dopamine in the NAcc in male and female prairie voles. Microinjection of a dopamine receptor antagonist into the NAcc prevents the formation of mating-induced partner preference, whereas treatment with a dopamine receptor agonist facilitates partner preference formation in the absence of mating. In addition, in females, hypothalamic Oxt signaling promotes dopamine release in the NAcc, and concurrent activation of Oxt receptors and DRD2 in the NAcc is essential for partner preference formation. Thus, mesolimbic dopamine release during social and sexual interactions may increase the salience of the rewarding aspects of interacting with a new social partner. Moreover, the distinct subgroups of dopamine receptors have opposite behavioral effects. Specifically, partner preference formation is facilitated by NAcc-targeted administration of a DRD2 agonist and blocked by a local injection of a DRD1 agonist in the NAcc in males. Intriguingly, pair-bonded males have a higher density of DRD1 in NAcc compared with sexually naive male prairie voles. As DRD1 activity prevents partner preference formation, this may be a neural mechanism that hinders pair-bonded males from establishing new social preferences or ascribing rewarding connotations to interactions with other conspecifics. Thus, the mesolimbic dopamine system is responsible for both partner preference formation and maintaining this selective preference in prairie voles.

SELECTIVE AGGRESSION

Sexually naive prairie voles are generally gregarious and display low levels of aggression to conspecifics. After 24 h of cohabitation and mating, male prairie voles display high levels of aggression toward strangers, particularly unfamiliar males, but not toward their female partner. Prolonged cohabitation, exceeding a week, primes males to display aggression to all intruding conspecifics to their territory, regardless of sex. Although mating-induced aggression is common in many species, this adaptive behavior is usually in the form of temporary mate guarding to improve reproductive success via paternal certainty. In contrast, the bond-induced aggression observed in prairie voles is equally directed to unfamiliar male and female conspecifics, averting rivals and sacrificing potential extra-pair mating opportunities. Furthermore, bonded female prairie voles display overt aggression to intruders, an extremely rare event across mammalian species. This is one behavioral mechanism through which pair bonds are preserved, and studies have indicated the involvement of both Avp and dopamine circuits in this behavior (Gobrogge & Wang, 2016).

Initial experiments noted that Avpr1a binding in the anterior hypothalamus (AH) is increased in males paired with a female. Furthermore, neuronal activity in the AH and medial amygdala (MeA), particularly the Avp neurons, is increased following aggressive confrontation with a stranger conspecific but silent after nonaggressive interactions with the female partner. Avp is released in the AH during aggressive confrontations with intruders. Pharmacology studies demonstrate this Avp release has a functional role in regulating aggressive behavior as local injections of Avp into the AH can increase aggressive behavior, whereas concurrent treatment with an Avpr1a antagonist eliminates the rise in intruder aggression. Recent in vivo behavioral pharmacology and real-time reverse microdialysis experiments demonstrated that Avp-induced activity in the AH and subsequent aggressive behavior in prairie voles are increased by corticotrophin-releasing hormone (CRH) signaling from the MeA and suppressed by serotonin (5-HT) signaling from the dorsal raphe (DR) (Gobrogge, Jia, Liu, & Wang, 2016). Intriguingly, the CRH$^{MeA \to AH}$ neurons are activated during encounters with intruding conspecifics, but silent during periods of interactions with a bonded partner. This pattern of activity is opposite for 5-HT$^{DR \to AH}$ neurons. Thus, a microcircuit exists in the pair-bonded male brain to switch the propensity for aggression or affiliation within the appropriate social context, propagating the maintenance of the pair bond.

The mesolimbic dopamine circuitry is also involved in selective aggression, particularly aggression displayed by pair-bonded males targeting strange females (Aragona & Wang, 2009). Unlike the generally gregarious and low agonistic nature of sexually naive males, males cohabitating and mating with a female for 24 h will display overt aggression toward strange males. Although these newly bonded males eventually display aggression toward strange females as well, it requires an extended cohabitation period (up to 2 weeks). Interestingly, although the DRD1 density in the NAcc is comparable between sexually naive males and those living and mating with a female for 24 h, it is significantly increased after 2 weeks of exposure to a female. This synchronized time course between the shift in DRD1 expression in the NAcc and aggression directed toward intruding females stimulated the exploration of DRD1 activity in the NAcc in regulation of this behavior. Pharmacology studies confirmed that antagonizing DRD1 function in the NAcc extinguished this form of aggression. Thus, DRD1 function is not involved in the early phase of pair bond formation (e.g., male-male aggression), given this neuroplastic reorganization does not occur during the first 24 h of cohabitation but rather over the next 2 weeks. However, these findings highlight its role in regulation of the full establishment of a pair bond. Together with the research documenting that DRD2 in the NAcc is involved in the formation of a partner preference, the restructuring of the DRD2:DRD1 ratio in the NAcc, from the upregulation of DRD1, provides a neural mechanism that shifts the male prairie vole brain from a state of high social tolerance and openness to new social pairings to one that features a selective preference for a single social partner and antagonism toward other conspecifics.

SEX DIFFERENCES IN THE ROLE OF STRESS ON PAIR BOND FORMATION

The hypothalamic-pituitary-adrenal (HPA) axis of the prairie vole is sensitive to the social environment (Table 7.2). Its involvement provides a mechanism through which environmental challenges and stressful life events impact the formation of a pair bond between male and female prairie voles. For example, natal dispersal leads to increased reproductive fitness in male and female prairie voles, as young voles experience reproductive suppression and limited access to reproductive, unrelated conspecifics while in the natal group (Solomon & Jacquot, 2002). Before establishing male-female pairings, dispersed voles may live in communal groups or experience social isolation as a solitary individual or "wanderer." Males are twice as likely to adopt a wandering strategy than females. Thus, male prairie voles may endure more stressful conditions prior to forming pair bonds than females and, therefore, derive reproductive benefits from forming such bonds under, and despite of, stressful conditions.

In the laboratory, it has been demonstrated that the activity of the HPA axis does in fact modulate the formation of pair bonds in a sex-specific manner. Corticosterone levels, the end product of the HPA axis, in naive females and males precipitously decline by approximately 50% after exposure to an unfamiliar, opposite-sex conspecific but not to an unfamiliar, same-sex conspecific (DeVries, 2002). Within this period, female prairie voles will form a strong, long-term social preference for a new male partner. The length of cohabitation with an unfamiliar male to form a partner

TABLE 7.2   Sex-Specific Response of the HPA Axis to the Social Environment

Sociality	Male	Female	References
Male-female pairing	↓	↓	DeVries (2002)
Social isolation	⇔	↑	Grippo, Wu, Hassan, and Carter (2008)
Partner loss	↑	↑	Bosch et al. (2009), Sun et al. (2014)
Social buffering	No data	↓	Smith and Wang (2014)
Observed distress of partner	↑	No data	Burkett et al. (2016)

preference is reduced after experimentally reducing corticosterone levels in females via adrenalectomy, whereas supplemental replacement of corticosterone after adrenalectomy prevents formation. Furthermore, stress or injections of corticosterone prior to cohabitation prevents a partner preference to form in female prairie voles. In contrast, males will accelerate partner preference formation under periods of stress or treatment with corticosterone. Pharmacology studies also documented that administering CRH, the catalyzing agent of the HPA axis, into the NAcc will facilitate partner preference formation, whereas concurrent treatment with a CRH receptor antagonist prevents such behavior in male prairie voles (Lim et al., 2007). This suggests that partner preference formation is modulated by stress and that the HPA axis provides a biological mechanism that differentially impacts such social behaviors in male and female prairie voles.

SOCIAL BUFFERING AND CONSOLING: REFLECTIONS OF EMPATHY

Pair bonds in prairie voles may also be preserved by the experience of separation anxiety, namely, the distress ascribed to social separation and the comfort during reunion with a social partner. The HPA axis of prairie voles is highly tuned to the close proximity of a social partner. As previously mentioned, both males and females experience a drop in circulating corticosterone concentrations during the initial cohabitation with an opposite-sex conspecific (DeVries, 2002). Once heterosexual pairs are formed, corticosterone levels rise shortly after separation from a partner. In fact, prolonged separation and, moreover, social loss can have profound effects on stress physiology and normal behavioral routines of prairie voles that will be broached in a later section of this chapter. Finally, reunion with a familiar partner is associated with a significant decline in corticosterone levels. The HPA axis response to social contact after separation from a social partner is not universal to all forms of social stimuli as contact with an unfamiliar conspecific results in elevated corticosterone levels, particularly compared with those in prairie voles reuniting with their familiar partner. Thus, prairie voles are incentivized to maintain established social partnerships to avoid the consequences of separation.

Furthermore, social bonds provide protective buffering from the negative consequences of stress. Pair-bonded females under duress will display a greater resiliency when their male partner is in close proximity. A few recent studies have documented behavioral and neural mechanisms through which male- and female-bonded partners console each other during periods of distress, which yields social buffering of the stress response. After an acute exposure to an intense psychological stressor (i.e., immobilization stress), female prairie voles will display more stress-related stereotypies in the home cage if recovering alone (Smith & Wang, 2014). However, if their male partner is present, such behavior is absent as is the rise in circulating corticosterone levels and anxiety-like behavior in response to subsequent stress. This social buffering effect by the male partner is accompanied by a release of Oxt into the PVN, the region that serves as the central hub for the HPA axis in the brain. Pharmacological experiments documented that Oxt receptor antagonism in the PVN blocks these beneficial social buffering effects. A local circuit exists in the PVN of female prairie voles in which Oxt augments the inhibitory tone of GABA interneurons on CRH neurons through $GABA_A$ receptors (Smith et al., 2016). This provides a neural mechanism in the distressed female underlying the protective effects of a male partner.

Intriguingly, during a period of distress, female prairie voles do not display active social coping strategies, such as seeking out the contact of their partner (Smith & Wang, 2014). Rather, it is the male partner that displays consolation behavior by increasing partner-directed contact and grooming. The enhanced grooming behavior by the male only occurs if the female partner experiences a stressor (i.e., light foot shocks) and not when she is simply separated from the male for a brief period of time (Burkett et al., 2016). Intriguingly, this prosocial response by male prairie voles may be empathy based as males who detect the stress state of their partner also display emotional contagion, physiological state matching, familiarity bias, and self-other differentiation, all characteristics of the empathy hypothesis. Specifically, male prairie voles will mimic the level of anxiety- and fear-related behaviors and have plasma corticosterone levels that positively correlate with those of their distress partner. This increased grooming response by the male is not reciprocated by the distressed partner and is only offered when the distressed individual is a familiar partner. Furthermore, this prosocial behavior is accompanied by increased activity in the anterior cingulate cortex, a region expressing Oxt receptor in the male prairie vole brain, and can be abolished by an infusion of a selective Oxt receptor antagonist in this region. Thus, Oxt signaling is responsible for both the consolation behavior displayed by male prairie voles to their distressed partner and the buffering effects that such social interaction produces in the distressed female.

PARTNER LOSS: DEPRESSION AND BOND DISRUPTION

Social living is advantageous to the health and reproductive fitness of prairie voles. When the stability of the social group is challenged, prairie voles display a heightened state of distress, both behaviorally and physiologically. This is particularly true when there is a disruption of the pair bond due to separation or death of the social partner.[1] Partner loss in prairie voles produces autonomic imbalance and increases the HPA basal tone and responsivity to acute stressors (McNeal et al., 2014). Concurrently, voles display increased heart rates, heart rhythm dysregulation, and increased anxiety- and depressive-like behaviors, occurring within the first 24 h and persisting for at least 4 weeks (Bosch, Nair, Ahern, Neumann, & Young, 2009; McNeal et al., 2014; Sun, Smith, Lei, Liu, & Wang, 2014). Despite the absence of the partner, the behavioral mechanisms of pair bonds, namely, partner preference and selective aggression, are maintained for up to a week and, in some cases, 2 weeks (Insel & Hulihan, 1995; Sun et al., 2014; Winslow, Hastings, Carter, Harbaugh, & Insel, 1993). Thus, partner loss results in an immediate and extended dysregulation of stress behavior and physiology and, ultimately, dissolution of the pair bond itself.

Recently, several features of the male prairie vole brain have been identified to be sensitive to partner loss and influence the consequences of such loss. First, partner loss decreases Oxt expression in the PVN and Oxt receptor binding in the NAcc during the first 5 days after losing a bonded partner (Bosch et al., 2016). After 4 weeks, Oxt expression in the PVN reverses and rises to levels higher than in males in stable pairings (Sun et al., 2014). Pharmacology studies demonstrate that Oxt infusion in the NAcc blocks the passive coping or depressive-like behaviors of social loss. In addition, Oxt release in the NAcc is manipulated by central infusion of CRH receptor 2 (CRHR2) ligands, increased by a CRHR2 antagonist, and decreased by a CRHR2 agonist. Second, social loss increases CRH expression in the BNST, a brain region with CRH neurons that project to the NAcc, and the consequences of loss can be blocked by pharmacological antagonism of CRHR2 in the NAcc. Together, this suggests a NAcc-centric mechanism where CRH and Oxt inputs have inversed regulation of the consequences of partner loss.

---

[1]Social isolation is coupled to the challenge of losing a bonded partner. Intriguingly, naive adult males, or males not exposed to an unrelated female conspecific, are relatively resilient to separation from a same-sex sibling (Grippo et al., 2008). In contrast, females experiencing isolating from a male partner or female sibling display similar behavioral and physiological distress. Although male-female pairs are the most common social group, single wanders and communal groups also exist. When evaluating the social groups of unpaired males and females, males are twice as likely to be wanders than females, communal groups include more females than males, and males disperse from communal groups sooner and are less likely to join new groups than females (Getz and McGuire, 1997; McGuire, Getz, Hofmann, Pizzuto, & Frase, 1993). These differences in the social strategies of unpaired adult male and female prairie voles may be a mechanism for the sex difference in response to social isolation.

FINAL REMARKS

This chapter highlights the utility of prairie voles to study the impact of stress on neurobiological and behavioral mechanisms underlying pair bond formation, maintenance, and disruption as well as consolation and social buffering. Prairie voles show remarkable behavioral and physiological responses to changes in the social environment, particularly as it pertains to disturbances to the pair bond. Undoubtedly, the HPA axis serves as a prominent biological mechanism to regulate such changes. Hormones of the HPA axis (e.g., CRH and corticosterone) impact the formation of pair bonds in a sex-specific manner, hindering bonding in females while facilitating it in males. In addition, consequences of HPA axis dysregulation that occurs after the loss of a social partner is a driving force underlying the maintenance and stability of this relationship. Data have also shown that Oxt, directly or by tempering CRH signaling, functions as neuromodulator of stress, pair bonding, and social interaction. During stressful life events, Oxt signaling in the brain of male and female prairie voles can promote the consolation of a distressed partner and elicits the social buffering effects by inhibiting stress-associated CRH signaling. In addition, CRH and Oxt inputs inversely regulate the consequences of partner loss. Finally, it could be speculated that dopamine and CRH signaling in the NAcc interact to impact different aspects of the pair bond status (from partner preference behavior in pair bond formation to depressive behavioral response resulting from partner loss). Ultimately, further prairie vole research focusing on the interactions between CRH, Oxt, and dopamine signaling within specific brain regions would provide a neural circuit that features the impact of stress and neuromodulatory pathways on the pair bond from formation through the consequences of its dissolution. Social living is common in prairie voles and has been shaped by environmental challenges. The introduction of the prairie vole model to laboratory has resulted in an ever-expanding body of research highlighting the neurobiology that is involved in adapting the vole sociality to such pressures.

Acknowledgments

Funding for this work was provided by National Institutes of Health Grants MHR01-58616 and MHR01-89852 to ZW.

DISCLOSURE STATEMENTS

The authors declare there are no conflicts to disclose. the authors are funded by NIMH & NIDA only. Grant Support: National Institutes of Health Grants MHR01-58616 and MHR01-89852 to ZW.

References

Aragona, B. J., & Wang, Z. (2009). Dopamine regulation of social choice in a monogamous rodent species. *Frontiers in Behavioral Neuroscience, 3*, 15.

Bosch, O. J., Dabrowska, J., Modi, M. E., Johnson, Z. V., Keebaugh, A. C., Barrett, C. E., & Young, L. J. (2016). Oxytocin in the nucleus accumbens shell reverses CRFR2-evoked passive stress-coping after partner loss in monogamous male prairie voles. *Psychoneuroendocrinology, 64*, 66–78.

Bosch, O. J., Nair, H. P., Ahern, T. H., Neumann, I. D., & Young, L. J. (2009). The CRF system mediates increased passive stress-coping behavior following the loss of a bonded partner in a monogamous rodent. *Neuroendocrinology, 34*, 1406–1415.

Bowlby, J. (1958). The nature of the child's tie to his mother. *International Journal of Psycho-Analysis, 39*, 350–373.

Burkett, J. P., Andari, E., Johnson, Z. V., Curry, D. C., de Waal, F. B. M., & Young, L. J. (2016). Oxytocin-dependent consolation behavior in rodents. *Science, 351*, 375–378.

Cho, M. M., DeVries, A. C., Williams, J. R., & Carter, C. S. (1999). The effects of oxytocin and vasopressin on partner preferences in male and female prairie voles (*Microtus ochrogaster*). *Behavioral Neuroscience, 113*, 1071–1080.

DeVries, A. C. (2002). Interaction among social environment, the hypothalamic-pituitary-adrenal axis, and behavior. *Hormones and Behavior, 41*, 405–413.

Getz, L. L., & McGuire, B. (1997). Communal nesting in prairie voles (*Microtus ochrogaster*): formation, composition, and persistence of communal groups. *Canadian Journal of Zoology, 75*, 525–534.

Gobrogge, K. L., Jia, X., Liu, Y., & Wang, Z. (2017). Neurochemical mediation of affiliation and aggression associated with pair bonding. *Biological Psychiatry, 81*(3), 231–242.

Gobrogge, K., & Wang, Z. (2016). The ties that bond: neurochemistry of attachment in voles. *Current Opinion in Neurobiology, 38*, 80–88.

Grippo, A. J., Wu, K. D., Hassan, I., & Carter, C. S. (2008). Social isolation in prairie voles induces behaviors relevant to negative affect: toward the development of a rodent model focused on co-occurring depression and anxiety. *Depress and Anxiety, 25*, E17–26.

Insel, T. R., & Hulihan, T. J. (1995). A gender specific mechanism for pair bonding: oxytocin and partner preference formation in monogamous voles. *Behavioral Neuroscience, 109*, 782–789.

Lim, M. M., Liu, Y., Ryabinin, A. E., Bai, Y., Wang, Z., & Young, L. J. (2007). CRF receptors in the nucleus accumbens modulate partner preference in prairie voles. *Hormones and Behavior, 51*, 508–515.

Lim, M. M., & Young, L. (2004). Vasopressin-dependent neural circuits underlying pair bond formation in the monogamous prairie vole. *Neuroscience, 125*, 35–45.

McGuire, B., Getz, L. L., Hofmann, J., Pizzuto, T., & Frase, B. (1993). Natal dispersal and philopatry in prairie voles (*Microtus ochrogaster*) in relation to population density, season, and natal social environment. *Behavioral Ecology and Sociobiology, 32*, 293–302.

McNeal, N., Scotti, M. A., Wardwell, J., Chandler, D. L., Bates, S. L., Larocca, M., & Grippo, A. J. (2014). Disruption of social bonds induces behavioral and physiological dysregulation in male and female prairie voles. *Autonomic Neuroscience, 180*, 9–16.

Neumann, I. D. (2009). The advantage of social living: brain neuropeptides mediate the beneficial consequences of sex and motherhood. *Frontiers in Neuroendocrinology, 30*, 483–496.

Ophir, A. G., Gessel, A., Zheng, D. J., & Phelps, S. M. (2012). Oxytocin receptor density is associated with male mating tactics and social monogamy. *Hormones and Behavior, 61*, 445–453.

Pizzuto, T., & Getz, L. L. (1998). Female prairie voles (*Microtus ochrogaster*) fail to form a new pair after loss of mate. *Behavioural Processes, 43*, 79–86.

Ross, H. E., & Young, L. J. (2009). Oxytocin and the neural mechanisms regulating social cognition and affiliative behavior. *Frontiers in Neuroendocrinology, 30*, 534–547.

Smith, A. S., Tabbaa, M., Lei, K., Eastham, P., Butler, M. J., Linton, L., & Wang, Z. (2016). Local oxytocin tempers anxiety by activating GABA receptors in the hypothalamic paraventricular nucleus. *Psychoneuroendocrinology, 63,* 50–58.

Smith, A. S., & Wang, Z. (2014). Hypothalamic oxytocin mediates social buffering of the stress response. *Biological Psychiatry, 76,* 281–288.

Solomon, N. G., & Jacquot, J. J. (2002). Characteristics of resident and wandering prairie voles *Microtus ochrogaster. Canadian Journal of Zoology, 80,* 951–955.

Sun, P., Smith, A. S., Lei, K., Liu, Y., & Wang, Z. (2014). Breaking bonds in male prairie vole: long-term effects on emotional and social behavior, physiology, and neurochemistry. *Behavioural Brain Research, 265,* 22–31.

Wang, H., Duclot, F., Liu, Y., Wang, Z. X., & Kabbaj, M. (2013). Histone deacetylase inhibitors facilitate partner preference formation in female prairie voles. *Nature Neuroscience, 16,* 919–924.

Wang, Z., Young, L. J., De Vries, G. J., & Insel, T. R. (1998). Voles and vasopressin: a review of molecular, cellular, and behavioral studies of pair bonding and paternal behaviors. *Progress in Brain Research, 119,* 483–499.

Williams, J. R., Catania, K. C., & Carter, C. S. (1992). Development of partner preferences in female prairie voles (*Microtus ochrogaster*): the role of social and sexual experience. *Hormones and Behavior, 26,* 339–349.

Winslow, J. T., Hastings, N., Carter, C. S., Harbaugh, C. R., & Insel, T. R. (1993). A role for central vasopressin in pair bonding in monogamous prairie voles. *Nature, 365,* 545–548.

Witt, D. M., Carter, C. S., & Insel, T. R. (1991). Oxytocin receptor binding in female prairie voles: endogenous and exogenous oestradiol stimulation. *Journal of Neuroendocrinology, 3,* 155–161.

The Social Transmission of Associative Fear in Rodents— Individual Differences in Fear Conditioning by Proxy

Carolyn E. Jones, Marie-H. Monfils
The University of Texas at Austin, Austin, TX, United States

DIRECT AND INDIRECT ASSOCIATIVE FEAR LEARNING

The innate abilities of animals to learn to recognize and respond to cues that predict danger stem from well-understood fear learning pathways (LeDoux, 1992; Maren, 2001). In Pavlovian fear conditioning, an initially neutral stimulus (e.g., tone, light, or context) is paired with an aversive unconditioned stimulus (US) (e.g., a footshock or tailshock) resulting in the formation of a fear memory such that later presentation of the now-conditioned stimulus (CS) in the absence of the US elicits a conditioned response (CR) (Pavlov, 1927) (Fig. 8.1A for design of Pavlovian fear conditioning). When rodents, who are small prey animals, are used as subjects, the CR typically measured as an index of fear is freezing. While freezing, the animal crouches and ceases any movement except those associated with respiration, presumably to avoid detection from a predator (Blanchard & Blanchard, 1969). In addition to the passive conditioned fear response of immobility, rats can also vocalize, attempt escape, or even fight, depending on the context/situation, in response to salient fear cues.

Learning through direct experience, however, is only one way through which individuals acquire aversive associations about their surroundings. For decades, researchers have investigated how primates, both human and nonhuman, indirectly learn about fear-inducing

Neuronal Correlates of Empathy. http://dx.doi.org/10.1016/B978-0-12-805397-3.00008-5

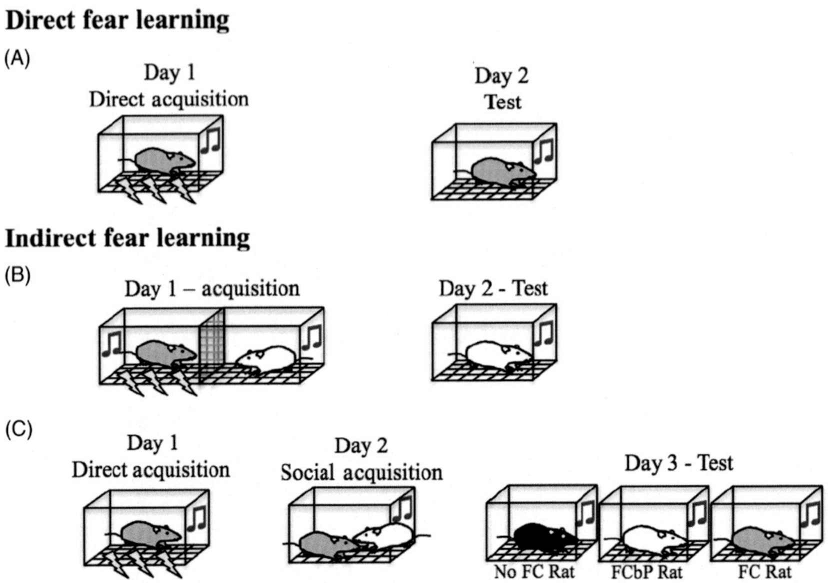

Direct fear learning

(A)

Indirect fear learning

(B)

(C)

FIGURE 8.1 **Direct and indirect fear conditioning designs.** (A) Direct fear learning usually involves Pavlovian fear conditioning. On Day 1, subjects are presented with a CS (sound) paired with an aversive US (shock). The following day, the CS is played alone and a fear response is measured. Fear display to the CS indicates a learned fear association. (B) Observer–demonstrator paradigms are the most commonly used methods for indirect fear learning. Subjects are in a modified fear-conditioning chamber, usually separated by a see-through physical barrier. An observer rat will watch as a demonstrator rat is FC with a CS–US pairing. A follow-up test on Day 2 to the CS alone is optional or not included at all in many observational fear paradigms. (C) In fear conditioning by proxy, rats are housed in triads. On Day 1, one rat of the triad is FC. On Day 2, the FC rat and a cagemate are returned to the chamber, and the CS is played. This session is called fear conditioning by proxy (FCbP), and rats can interact freely. On Day 3, long-term memory is tested by placing each rat in the chamber individually and presenting the CS.

stimuli through observation of a conspecific (Cook & Mineka, 1987; Cook, Mineka, Wolkenstein, & Laitsch, 1985; Hygge & Öhman, 1978; Mineka & Cook, 1988; Mineka, Davidson, Cook, & Keir, 1984; Olsson & Phelps, 2004). Still, research is just beginning in other mammals, including the laboratory rodent (Atsak et al., 2011; Bredy & Barad, 2009; Bruchey, Jones, & Monfils, 2010; Guzmán et al., 2009; Jeon et al., 2010; Kavaliers, Colwell, & Choleris, 2005; Kim, Kim, Covey, & Kim, 2010; Knapska, Mikosz, Werka, & Maren, 2010; Langford et al., 2006; Masuda, Aou, & Tsien, 2009; Pereira, Cruz, Lima, & Moita, 2012).

There is strong empirical support for the idea that laboratory rodents are influenced by the emotional states of others. The presence of a nonfearful conspecific greatly reduces fear expression when rodents are placed in either a novel (Guzmán et al., 2009) or a fear-conditioned (FC) context (Kiyokawa, Kikusui, Takeuchi, & Mori, 2004). Conversely, the presence of a fearful rat increases freezing in a FC animal (Atsak et al., 2011). After interacting with a FC cagemate, an otherwise naive rat shows increases in amygdala activity (as evidenced by increased c-Fos labeling; Knapska et al., 2006) interacts differently with its cagemates and shows enhanced fear learning (Knapska et al., 2010). When previously naive rats interact with a cagemate behaving fearfully in the presence of a CS (in the absence of any US), subsequent fear conditioning to a subthreshold US is heightened compared to rats that were not exposed to fearful rats (Bruchey et al., 2010).

Together, these findings demonstrate that (1) the emotional state of one rodent can influence the behavior and neural activity of another, and (2) rodents can learn a fear association in the absence of direct experience with a US. This has lead to a surge in new research exploring various methods of indirect fear learning where one animal observes another demonstrate fear behavior. Since the early 2000s, paradigms designed to investigate indirect fear conditioning to previously neutral cues in laboratory rodents are rapidly emerging in the literature, but variations in the design and execution of these paradigms allow for drastically different interpretations of the results.

Design differences in social fear-learning paradigms allow researchers to approach the indirect transmission of associative fear from several angles. The development of new observational and social fear conditioning models is essential to understand the boundaries of social fear learning in rats and its relevance to vicarious fear learning and empathy in humans. In the majority of existing paradigms, design requirements inherent to the use of rodents as a research model (e.g., physical barriers between animals, as well as unknown social status/relations between observers and demonstrators) hinder the translational relevance of most observational fear-learning paradigms as means of investigating mechanisms of socially acquired fear and vicarious emotional responding in humans (Fig. 8.1B, for example, of observer–demonstrator social fear-conditioning paradigm).

FEAR CONDITIONING BY PROXY SOCIAL LEARNING PARADIGM

In an attempt to examine whether fear could be socially transmitted in rats during retrieval of a discrete memory, but in the absence of an immediate danger, we developed a paradigm we refer to as "fear conditioning by proxy" (Bruchey et al., 2010; Jones, Riha, Gore, & Monfils, 2014). Specifically, we asked whether a specific cue (a tone) could come to elicit fear expression in a rat after simply interacting with a conspecific, and observing their CR in the presence of this otherwise benign stimulus without the presence of a physical barrier.

This fear conditioning by proxy (FCbP) paradigm always makes use of rats housed in triads. On Day 1, one rat of each triad is FC to a cue paired with a footshock. On Day 2, the fear-conditioned rat (FC rat) is returned to the fear-conditioning chamber accompanied by a cagemate (FCbP rat), and the tone is played in the absence of the footshock. The third rat (No FC) remains in the home cage and on Day 2 and is allowed to freely interact with the FC and FC by proxy (FCbP) rat when they are returned to the colony after the CS presentations. The following day (Day 3), all rats (FC, FCbP, and No FC) are placed in the chambers alone and tested for fear expression (freezing) to the CS (Fig. 8.1C for fear conditioning by proxy experimental design). Testing in the absence of the demonstrator is essential to determine if learning has occurred and to tease out any emotional, motivational, or social facilitatory effects that can occur when animals are present in the same chamber.

When tested for freezing to the CS on Day 3, we consistently find a subset of rats that freeze significantly more than naive animals after exposure to a conspecific expressing fear to a conditioned cue. However, approximately 50% of FCbP rats do not freeze at all (Fig. 8.2 for dispersion of freezing at long-term memory test in fear conditioning by proxy paradigm) resulting in a large amount of individual variation in social fear acquisition through fear conditioning by proxy.

INDIVIDUAL FACTORS THAT CONTRIBUTE TO FEAR CONDITIONING BY PROXY

The modest effect of the FCbP paradigm, as well as the wide dispersal of behavior, combined with results from other laboratories indicating that naive rats do not display fear responses during observational fear-learning paradigms (Atsak et al., 2011; Pereira et al., 2012), underscores the importance of disambiguating factors that contribute to individual differences in social learning. Here we review some of the individual factors that contribute to fear conditioning by proxy, including social relationships between animals (both familiarity/kinship and social dominance structure), factors present during learning (vocalizations, physical contact between animals, and freezing displayed by the demonstrator rat), and previous experience of the social learner. We will discuss, in turn, familiarity/kinship, vocalizations, dominance hierarchy, demonstrator freezing, and previous fear experience.

Familiarity/Kinship

In observational fear-learning experiments where one animal observes a fear reaction in a conspecific, socially familiar or genetically related animals

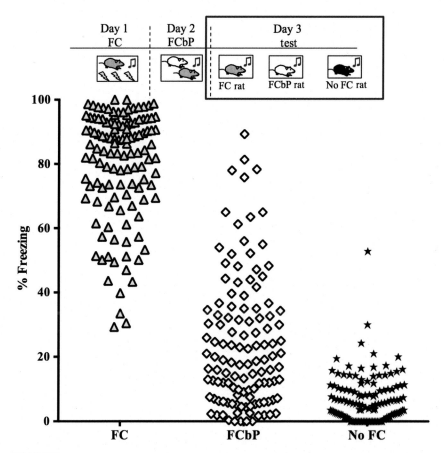

FIGURE 8.2 **Dispersion of freezing during long-term memory tests in fear conditioning by proxy social learning paradigm.** Directly FC rats typically freeze a high amount to CS presentation during long-term memory tests (*gray triangles*), and rats only exposed to fearful rats in the home cage (No FC) freeze very little to the CS (*black stars*). However, FCbP rats consistently show a wide range of freezing when tested to the cue 24 h after interacting with a fearful conspecific when a CS is played (*white diamonds*). *Source: Figure modified from Jones, C. (2015). The social transmission of associative fear in rats: Mechanisms and applications of fear conditioning by proxy (dissertation). The University of Texas at Austin, Texas Dissertation Libraries, Texas Dissertation Libraries database and represents all fear conditioning by proxy experiments performed in the Monfils laboratory.*

transfer greater degrees of response information (Jeon et al., 2010; Kavaliers et al., 2005; Kim et al., 2010; Pereira et al., 2012). Jeon et al. (2010) performed an elegant set of experiments where a mouse that observed another mouse receiving shock in a context showed fear when tested later in the same context. They found that mice with social relations (10+ weeks as a mating pair or siblings who grew up together) conditioned more strongly, and that the response was driven largely (but not entirely) by visual access to

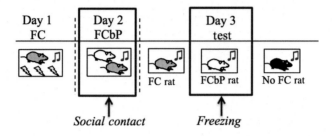

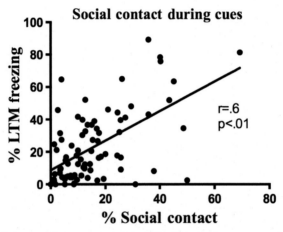

FIGURE 8.3   **Social contact during fear conditioning by proxy predicts fear display on the following day.** The percent of time spent engaged in social contact with the FC rat during CS presentation on Day 2 of the fear conditioning by proxy paradigm was positively correlated with the freezing displayed to the CS during long-term memory tests 24 h later.

the conditioning procedure (Jeon et al., 2010). In deer mice, both familiarity and genetic relatedness increased observational fear responses to biting flies (Kavaliers et al., 2005); however, neither of the demonstrator/observer pairs in the aforementioned experiments were allowed to physically interact with each other. With the FCbP design, rats freely interact during social learning. We demonstrate not only that Sprague Dawley rats acquire more fear information about a conditioned cue from a familiar and related conspecific, but also that social interactions occurring between pairs during fear conditioning by proxy modulate the degree of freezing during test (Fig. 8.3) (Bruchey et al., 2010; Jones et al., 2014).

Auditory Stimuli—Ultrasonic Vocalizations

In addition to visual cues, other groups have indicated that auditory information is the essential element of social transmission in fear paradigms, with the presence of negative affect vocalizations (Kim et al., 2010), or the sudden onset of silence (Pereira et al., 2012) as necessary indicators of danger in a social setting.

Vocalizations in the lower spectrum of the ultrasonic range (around 22 kHz) are associated with negative affect elicited in situations wherein the rat is fearful, such as the presence of a predator (Blanchard, Blanchard, Agullana, & Weiss, 1991), painful (Calvino, Besson, Boehrer, & Depaulis, 1996; Han, Bird, Li, Jones, & Neugebauer, 2005), or startling (Kaltwasser, 1991) stimuli, or in aggressive encounters with other rats (Thomas, Takahashi, & Barfield, 1983). These vocalizations usually consist of long, loud, monotone calls that show little variation in pitch. Higher-frequency vocalizations (referred to typically as 50-kHz calls, although it is not uncommon to be as high as 80–100 kHz) are affiliated with activities with more positive affect such as play (Knutson, Burgdorf, & Panksepp, 1998), exploration (Wöhr & Schwarting, 2007), and anticipation of reward (Burgdorf, Knutson, & Panksepp, 2000). Importantly, there are individual differences in vocalization profiles between animals (Ahrens et al., 2013).

We found that in the fear conditioning by proxy paradigm, the majority of rats do not vocalize at all during the social transmission of fear on Day 2, but this does not preclude a conspecific from learning about associative fear. In line with the results of Kim et al. (2010), however, we also found that for the rats that *did* vocalize in the 22-kHz range during fear conditioning by proxy on Day 2, the duration of those vocalizations was positively correlated with the amount of fear acquired socially, as measured by freezing displayed by the observer the following day.

Social relationships between rats in a cage, three weeks before FCbP, were categorized by observing samples of play within a cage after a 24-hour isolation (Jones & Monfils, 2016; Pellis & Pellis, 1991; Pellis, Pellis, & McKenna, 1993 for a detailed description of employed methods). This model suggested that the social relationship between rats, as measured by play behaviors (e.g., nape contacts, counter attacks, evasion, and pinning), may explain why some animals vocalize in response to threat in the presence of conspecifics and some do not, and further supports the idea that one function of 22-kHz ultrasonic vocalizations is to alarm conspecifics (Blanchard et al., 1991; Brudzynski & Chiu, 1995; Sales, 1991). The social relationship between rats may determine if and how alarm calls are used in threat-provoking situations (Blanchard et al., 1991; Blumstein & Armitage, 1997; Clutton-Brock et al., 1999). This may represent an important aspect to control for in laboratory experiments and increase translational relevance of behavioral paradigms in rats.

Dominance Hierarchy

Dominance hierarchies in wild animals that live in a group tend to emerge out of necessity as a way for animals to coexist in an environment where threats are constant and resources (e.g., food, water, mates, and shelter) are limited. As a group, members have access to more resources than they could obtain as an individual and can better defend their territory from threat.

In multiple species, including the rat, alpha males tend to take responsibility for initiating offensive action against potential threats (Blanchard, Fukunaga-Stinson, Takahashi, Flannelly, & Blanchard, 1984; Blanchard, Blanchard, Takahashi, & Kelley, 1977) and are allowed first access to food and mates. In the social learning paradigms that have controlled for social dominance (Jones & Monfils, 2016; Kavaliers et al., 2005), subordinate animals displayed more pronounced fear responses after a social learning session with a dominant demonstrator. By categorizing rats based on details of their offensive and defensive responses to play fighting (Pellis et al., 1993), we found that subordinate rats freeze more after fear conditioning by proxy with a dominant cagemate (Fig. 8.4) (Jones & Monfils, 2016). Together these results strongly support the importance of the social dominance relationship in the social transmission of fear.

Demonstrator Freezing

Freezing is a natural response to threat proximity in the rodent. As such, freezing displayed by the observer rat during the FCbP session may

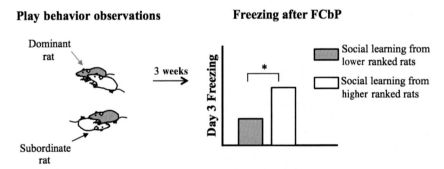

FIGURE 8.4 **Social dominance hierarchy influences learning by proxy.** When rats are classified into a dominance hierarchy by observing the frequency of dominant (e.g., counterattacks, nape contacts) or subordinate (e.g., rotating to supine) play behaviors, we find that rats freeze more after social fear learning with a higher-ranked cagemate than with a lower-ranked rat. Subordinate rats (white) froze more after fear conditioning by proxy with the dominant rat (gray).

not indicate a learned fear behavior but may instead be a response facilitated by the presence of a freezing conspecific. However, across multiple experiments that make use of the FCbP paradigm, we have consistently found that naive rats do not freeze in response to a fearful conspecific during fear conditioning by proxy, and the amount of freezing displayed by the fearful demonstrator is not correlated with the fear response displayed the following day in social learners (Fig. 8.5). This supports the idea that freezing behavior in one rat does not function the same as a higher-order CS (where increased intensity produces increased conditioned responding).

Many social learning theories insist that demonstrator outcome is essential for learning to occur vicariously. Evolutionarily, this seems intuitive: relying on a social learning strategy over direct experience or trial and error learning is less costly in terms of immediate risk and energy expenditure (the social learner does not have to risk threat exposure directly to

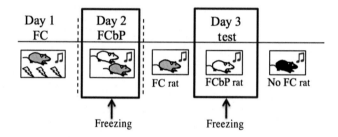

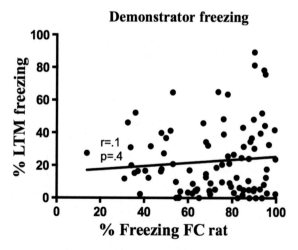

FIGURE 8.5 **Freezing displayed by the demonstrator FC rat during social learning is not correlated to freezing in the social learner FCbP rat. FC rat freezing on Day 2 is not correlated with FCbP rat freezing on Day 3.**

benefit from the learned aversion). Conversely, copying behavior of failed attempts with no regard for the situational relevance or potential effectiveness in producing a desired outcome could result in the spread of maladaptive behavior, thereby putting colony members, as well as oneself, at a disadvantage. When considering the multiday design of the fear conditioning by proxy paradigm, it could be reasoned that FCbP rats observed FC rats freezing to a novel cue (with social contact between animals a possible indicator of attention to or involvement in this situation), after which point they were both removed from the chamber, and there was no negative outcome. This would constitute a successful performance of the target behavior. When the FCbP rat does not attend socially to the FC rat, one possibility is that the observer rat may simply miss the behavior display of the FC rat, and at the end of the session both rats are still removed (the session ends regardless of subject performance). These rats have no reason to engage in a social learning strategy, because at this point their direct experience indicates that they will be removed after cue presentation no matter how they behave. This is congruent with the idea that rats with no previous direct experience with fear-inducing stimuli (footshock, stressor, or fearful conspecific) are less likely to display a fear response to a neutral cue.

Previous Fear Experience

In line with similar work (Atsak et al., 2011; Chen, Panksepp, & Lahvis, 2009; Davitz & Mason, 1955; Kim et al., 2010; Pereira et al., 2012), during Day 2 of the FCbP paradigm, we found that previously naive rats do not display any freezing while interacting with a fearful cagemate (Bruchey et al., 2010; Jones et al., 2014). One important difference between the FCbP paradigm presented here and other observational learning paradigms is that the FCbP rats are tested again 24 h after social fear acquisition. The exclusion of a follow-up test for retention of this socially transmitted fear makes it challenging to differentiate observed behavior as an emotional response to a conspecific in distress or the acquisition of a fear memory to a social stimulus.

By measuring freezing during long-term memory (LTM) tests the next day, we were able to examine retention of a fear memory after social acquisition and found that a subset of FCbP rats froze in response to the cue on Day 3. It may be that the fear responses of the FC demonstrator rat are not inherently aversive to the FCbP observer rat, because a fear response is not prompted during the acquisition session. This idea is further supported by a lack of correlation between freezing displayed by the FC rat and the freezing displayed during LTM by the FCbP observer. It may also be that desire to engage the FC rat socially overwhelms the desire to freeze when a conspecific is present.

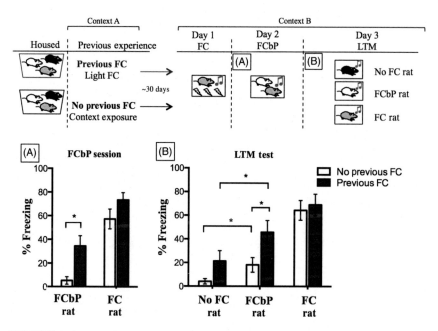

FIGURE 8.6 **Prior fear experience influences social fear acquisition.** Fear-experienced rats freeze more during both fear conditioning by proxy (A) and test the next day (B). *Source: Figure modified from Jones, C. (2015). The social transmission of associative fear in rats: Mechanisms and applications of fear conditioning by proxy (dissertation). The University of Texas at Austin, Texas Dissertation Libraries, Texas Dissertation Libraries database.*

In contrast, rats with prior fear experience show increased fear expression in response to a novel cue after (and during) fear conditioning by proxy. When rats were FC to a different CS (in a different sensory modality) one month prior to fear conditioning by proxy with a related and familiar cagemate, we found that rats with fear experience displayed more fear responding both during social fear exposure (FCbP) (Fig. 8.6A) and during follow-up tests 24 h later than naive animals (Fig. 8.6B). The effect of previous fear experience on fear conditioning by proxy (Jones, 2015; Jones & Monfils, 2016) adds to the substantial evidence accumulated in many other social and observational fear-learning paradigms establishing that both rats and mice acquire more fear socially if they have their own prior fear experiences to draw upon.

Rodents can learn associative fear from a conspecific; however, the conditions that facilitate this transfer of information vary and create a complex picture of the social world of the laboratory rodent with everything from social factors that exist within a cage and the relationship between observer and demonstrator to previous experience of the animal to the cues present during learning, both auditory and social (Fig. 8.7).

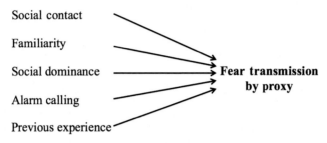

FIGURE 8.7   **Individual factors that contribute to social fear learning in the fear conditioning by proxy paradigm.** Social contact during fear conditioning by proxy, familiarity between rats, social dominance hierarchy, duration of 22-kHz alarm calls, and previous fear experience all contribute to the amount of socially learned fear in the fear conditioning by proxy paradigm.

SOCIAL FEAR LEARNING IN RODENTS—EVIDENCE FOR EMPATHY?

It is unmistakable that the emotional expressions of one rodent can influence both behavioral and physiological responses in another; however, whether these rodents are recognizing certain behavioral traits and matching the emotional state of another is unclear and may require a Theory of Mind as of yet unproven in rodents. Without asking subjects what they feel, extrapolating animal work on vicarious fear paradigms onto human empathy may be tenuous. Differences in social fear-learning paradigms make combining results difficult to interpret, and a blanket interpretation of this social fear transmission as evidence of empathy may be misguided. Instead, we look at what species-specific cues are necessary for social fear transmission. Do some of these cues convey threat while some invoke empathy?

Survival of any animal relies on its ability to associate environmental cues with the biologically relevant event they may predict. When these cues represent especially salient occurrences, such as the presence of food or a predator, it seems advantageous to the fitness of the species that the direct experiences of one animal can benefit, at least to some degree, other animals in the group.

It is interesting to note that the fear conditioning by proxy paradigm described here consistently reveals a subset of rats that do not appear to learn fear indirectly (as evidenced by a complete lack of freezing on long-term memory tests on Day 3), a phenomenon rarely seen in direct Pavlovian fear conditioning. The factors that determine these individual differences are the subject of further research but may be the result of differences of individual roles/relationships in the colony (or in this case, the cage; Blanchard, Flannelly, & Blanchard, 1988; Kavaliers et al., 2005; Shettleworth, 2010) resulting in differences in either social interactions or social learning. Differences in prior fear experience and sensory cues present also seem to be mitigating factors in social fear learning.

Learning that is socially facilitated is oftentimes attributed to stimulus enhancement, a phenomenon that occurs when a conspecific draws attention to, or enhances, the stimulus used for testing. Responding in the observer thereby occurs at a higher rate simply because his/her attention was diverted to a specific response. Observational conditioning (Mineka & Cook, 1988; 1993) can combine with stimulus enhancement to produce the fear response observed after fear conditioning by proxy. The fact that the target behavior (freezing) occurs in the absence of the conspecific when tested the next day is congruent with the idea that associative learning has occurred. Although if the observer rat were truly learning to fear the cues one would also expect to see this behavior during acquisition. It would appear that the presence of the demonstrator impedes the behavioral response (freezing) when fear conditioning has not previously occurred.

Importantly, it is unclear what the rat has learned to associate with the CS (e.g., what is the expected US). There could be an abstract possibility of danger inferred by the fear expressed in the demonstrator rat, or it could simply be that the fearful rat was aversive or threatening and thereby later presentation of the CS in the absence of the rat elicits a fear response in the observer in expectation of a fearful rat and consequently an agonistic encounter. The necessity of familiarity then may emerge as simply a contributing factor in the observer rat's ability to recognize the threat behavior of the conspecific. With either interpretation of social fear conditioning paradigms, increased social relevance of the demonstrator appears to be essential for detecting subtleties in his or her behavior directed at the CS, thereby drawing the attention of the observer to both the behavior of the demonstrator and the physical and sensory properties of the conditioned cues.

An alternate interpretation is that the social contact directed toward the FC rat while freezing is an attempt to reduce the fear response of the FC rat and indeed an indicator of a correctly identified emotional state that produces a prosocial behavioral response (e.g., consolation), an interpretation that aligns in some regards with definitions of empathy. If the purpose of such social contact is in fact to alleviate fear in the demonstrator, it is impossible to know the motive (e.g., selfish, to avoid conflict or altruistic, or to comfort another) behind such behavior. A major drawback of nonhuman animal research is that intent is extremely difficult to infer. If the social contact is an attempt to reduce the fear response, it is just as likely an attempt to avoid a conflict that may arise in an aroused animal as a means of self-protection (Koski & Sterck, 2009; Fraser & Bugnyar, 2010), and any conclusions regarding intent and motive (self vs. others) require extensive additional research.

Especially evident in prey animals is that there is an evolutionary advantage to living in groups and acting in ways that may benefit other members of the group. This does not necessarily provide evidence of empathy or compassion. Here, and in previous research, detection of a fear response (e.g., ultrasonic alarm calls; Kim et al., 2010), conditioned freezing or unconditioned startle response to a shock (Jeon et al., 2010), or lack of sounds that indicate normal movement (Pereira et al., 2012) in a conspecific may signal an immediate danger, and survival of the observer could depend on replicating that behavior. Disentangling a self-serving adaptive response to behavior displayed by a conspecific from the ability to detect and express shared emotions requires assumptions about intentionality and state of mind of the laboratory animal, which is not currently possible. Nonetheless, the behaviors observed in the FCbP paradigm described here along with multiple other paradigms that investigate social fear transmission mirror many elements consistent with empathy expression, including matched emotional behavior (e.g., freezing) and activation of brain regions in line with human empathic experiences (e.g., anterior cingulate cortex; Jones & Monfils, 2016).

CONCLUSIONS

To preserve Pavlov's initial terminology (Pavlov, 1927), when there is no "reflex" to respond with, a situation must be evaluated to determine the proper response, and this constitutes the main distinction between direct and indirect learning paradigms, or social learning paradigms. In socially relevant demonstrator rats, behavioral cues, such as those exhibited in response to a feared stimulus, may be especially salient thereby eliciting increased attention from the observer. When the behavioral cues of the demonstrator are not salient enough to draw the attention of the observer, the default response in the situation is reflexive learning based on previous direct exposure to the cues and thereby is a "no fear" behavioral response.

Although the fear display of the demonstrator does not seem to be aversive enough to induce a fear response during social fear acquisition, increased social attention to the fearful animal, as indicated by frequent contact between the two rats, consistently predicts subsequent fear behavior when the conditioned cue is played in isolation. If we assume that when a higher-order conditioned cue is presented (in contrast to a natural predator) a rat must make a judgment of "threat" or "no threat" along with an estimate of the magnitude of such threat that presumably determines the behavioral response, when the only previous experience with this cue was in the presence of a socially relevant rat behaving fearfully, the animal will tend to display a fear response as well, although lack of knowledge about the magnitude of the impending danger is likely responsible for the only moderate fear levels displayed.

In this chapter, we reviewed factors that influence the social transmission of associative fear in observational fear conditioning and the modified fear conditioning by proxy designs. This research provides perspective within the field as a whole and helps us to better understand both social and asocial learning strategies and the contingencies that may encourage the use of one strategy over another.

References

Ahrens, A. M., Nobile, C. W., Page, L. E., Maier, E. Y., Duvauchelle, C. L., & Schallert, T. (2013). Individual differences in the conditioned and unconditioned rat 50-kHz ultrasonic vocalizations elicited by repeated amphetamine exposure. *Psychopharmacology*, *229*(4), 687–700.

Atsak, P., Orre, M., Bakker, P., Cerliani, L., Roozendaal, B., Gazzola, V., et al. (2011). Experience modulates vicarious freezing in rats: A model for empathy. *PLoS One*, *6*(7), e21855–e121855.

Blanchard, D. C., Fukunaga-Stinson, C., Takahashi, L. K., Flannelly, K. J., & Blanchard, R. J. (1984). Dominance and aggression in social groups of male and female rats. *Behavioural Processes*, *9*(1), 31–48.

Blanchard, R. J., Flannelly, K. J., & Blanchard, D. C. (1988). Life-span studies of dominance and aggression in established colonies of laboratory rats. *Physiology & Behavior*, *43*, 1–7.

Blanchard, R., & Blanchard, D. (1969). Crouching as an index of fear. *Journal of Comparative and Physiological Psychology*, *67*(3), 370–375.

Blanchard, R. J., Blanchard, D. C., Agullana, R., & Weiss, S. M. (1991). Twenty-two kHz alarm cries to presentation of a predator, by laboratory rats living in visible burrow systems. *Physiology & Behavior*, *50*(5), 967–972.

Blanchard, R. J., Blanchard, D. C., Takahashi, T., & Kelley, M. J. (1977). Attack and defensive behaviour in the albino rat. *Animal Behaviour*, *25*, 622–634.

Blumstein, D. T., & Armitage, K. B. (1997). Alarm calling in yellow-bellied marmots: I. The meaning of situationally variable alarm calls. *Animal Behaviour*, *53*(1), 143–171.

Bredy, T. W., & Barad, M. (2009). Social modulation of associative fear learning by pheromone communication. *Learning & Memory*, *16*(1), 12–18.

Bruchey, A. K., Jones, C. E., & Monfils, M. -H. (2010). Fear conditioning by-proxy: Social transmission of fear during memory retrieval. *Behavioural Brain Research*, *214*(1), 80–84.

Brudzynski, S. M., & Chiu, E. M. (1995). Behavioural responses of laboratory rats to playback of 22 kHz ultrasonic calls. *Physiology & Behavior*, *57*(6), 1039–1044.

Burgdorf, J., Knutson, B., & Panksepp, J. (2000). Anticipation of rewarding electrical brain stimulation evokes ultrasonic vocalization in rats. *Behavioral Neuroscience*, *114*(2), 320–327.

Calvino, B., Besson, J. M., Boehrer, A., & Depaulis, A. (1996). Ultrasonic vocalization (22-28 kHz) in a model of chronic pain, the arthritic rat: Effects of analgesic drugs. *NeuroReport*, *7*(2), 581–584.

Chen, Q., Panksepp, J. B., & Lahvis, G. P. (2009). Empathy is moderated by genetic background in mice. *PLoS One*, *4*(2), e4387.

Clutton-Brock, T. H., O'Riain, M., Brotherton, P., Gaynor, D., Kansky, R., Griffin, A., et al. (1999). Selfish sentinels in cooperative mammals. *Science*, *284*(5420), 1640–1644.

Cook, M., & Mineka, S. (1987). Second-order conditioning and overshadowing in the observational conditioning of fear in monkeys. *Behaviour Research and Therapy*, *25*(5), 349–364.

Cook, M., Mineka, S., Wolkenstein, B., & Laitsch, K. (1985). Observational conditioning of snake fear in unrelated rhesus monkeys. *Journal of Abnormal Psychology*, *94*(4), 591.

Davitz, J. R., & Mason, D. J. (1955). Socially facilitated reduction of a fear response in rats. *Journal of Comparative and Physiological Psychology*, *48*(3), 149–151.

Fraser, O. N., & Bugnyar, T. (2010). Do ravens show consolation? Responses to distressed others. *PLoS One, 5*, e10605.

Guzmán, Y. F., Tronson, N. C., Guedea, A., Huh, K. H., Gao, C., & Radulovic, J. (2009). Social modeling of conditioned fear in mice by non-fearful conspecifics. *Behavioural Brain Research, 201*(1), 173–178.

Han, J. S., Bird, G. C., Li, W., Jones, J., & Neugebauer, V. (2005). Computerized analysis of audible and ultrasonic vocalizations of rats as a standardized measure of pain-related behavior. *Journal of Neuroscience Methods, 141*(2), 261–269.

Hygge, S., & Öhman, A. (1978). Modeling processes in the acquisition of fears: Vicarious electrodermal conditioning to fear-relevant stimuli. *Journal of Personality and Social Psychology, 36*(3), 271–279.

Jeon, D., Kim, S., Chetana, M., Jo, D., Ruley, H. E., Lin, S. -Y., et al. (2010). Observational fear learning involves affective pain system and Cav1.2 Ca2+ channels in ACC. *Nature Neuroscience, 13*(4), 482–488.

Jones, C. (2015). *The social transmission of associative fear in rats: Mechanisms and applications of fear conditioning by proxy (dissertation).* The University of Texas at Austin, Texas Dissertation Libraries. Texas Dissertation Libraries database.

Jones, C. E., & Monfils, M. H. (2016). Dominance status predicts social fear transmission in laboratory rats. *Animal Cognition.* Available at: http://doi.org/10.1007/s10071-016-1013-2.

Jones, C. E., & Monfils, M. H. (2016). Post-retrieval extinction in adolescence prevents return of juvenile fear. *Learning and Memory, 23*, 567–575.

Jones, C. E., Riha, P. D., Gore, A. C., & Monfils, M. H. (2014). Social transmission of Pavlovian fear: Fear-conditioning by-proxy in related female rats. *Animal Cognition, 17*(3), 827–834. doi: 10.1007/s10071-013-0711-2.

Kaltwasser, M. T. (1991). Acoustic signaling in the black rat (Rattus rattus). *Journal of Comparative Psychology, 104*, 227–232.

Kavaliers, M., Colwell, D. D., & Choleris, E. (2005). Kinship, familiarity and social status modulate social learning about "micropredators"(biting flies) in deer mice. *Behavioral Ecology and Sociobiology, 58*(1), 60–71.

Kim, E. J., Kim, E. S., Covey, E., & Kim, J. J. (2010). Social transmission of fear in rats: The role of 22-kHz ultrasonic distress vocalization. *PLoS One, 5*(12), e15077.

Kiyokawa, Y., Kikusui, T., Takeuchi, Y., & Mori, Y. (2004). Partner's stress status influences social buffering effects in rats. *Behavioral Neuroscience, 118*, 798–804.

Knapska, E., Nikolaev, E., Boguszewski, P., Walasek, G., Blaszczyk, J., & Kaczmarek, L. (2006). Between-subject transfer of emotional information evokes specific pattern of amygdala activation. *Proceedings of the National Academy of Sciences of the United States of America, 103*, 3858–3862.

Knapska, E., Mikosz, M., Werka, T., & Maren, S. (2010). Social modulation of learning in rats. *Learning & Memory, 17*(1), 35–42.

Knutson, B., Burgdorf, J., & Panksepp, J. (1998). Anticipation of play elicits high- frequency ultrasonic vocalizations in young rats. *Journal of Comparative Psychology, 112*(1), 65–73.

Koski, S. E., & Sterck, E. H. (2009). Post-conflict third-party affiliation in chimpanzees: what's in it for the third party? *American Journal of Primatology, 71*, 409–418.

Langford, D. J., Crager, S. E., Shehzad, Z., Smith, S. B., Sotocinal, S. G., Levenstadt, J. S., et al. (2006). Social modulation of pain as evidence for empathy in mice. *Science, 312*(5782), 1967–1970.

LeDoux, J. E. (1992). Brain mechanisms of emotion and emotional learning. *Current Opinion in Neurobiology, 2*(2), 191–197.

Maren, S. (2001). Neurobiology of Pavlovian fear conditioning. *Annual Review of Neuroscience, 24*(1), 897–931.

Masuda, A., Aou, S., & Tsien, J. (2009). Social transmission of avoidance behavior under situational change in learned and unlearned rats. *PLoS One, 4*(8), e6794.

Mineka, S., & Cook, M. (1988). Social learning and the acquisition of snake fear in monkeys. In *Social learning: Psychological and biological perspectives* (pp. 51–73). .

Mineka, S., & Cook, M. (1993). Mechanisms involved in the observational conditioning of fear. *Journal of Experimental Psychology: General, 122*(1), 23–38.

Mineka, S., Davidson, M., Cook, M., & Keir, R. (1984). Observational conditioning of snake fear in rhesus monkeys. *Journal of Abnormal Psychology, 93*(4), 355–372.

Olsson, A., & Phelps, E. A. (2004). Learned fear of "unseen" faces after Pavlovian, observational, and instructed fear. *Psychological Science, 15*(12), 822–828.

Pavlov, I. P. (1927). *Conditioned reflexes: An investigation of the physiological activity of the cerebral cortex.* London: Oxford University Press.

Pellis, S. M., & Pellis, V. C. (1991). Role reversal changes during the ontogeny of play fighting in male rats: Attack vs. defense. *Aggressive Behavior, 17*(3), 179–189.

Pellis, S. M., Pellis, V. C., & McKenna, M. M. (1993). Some subordinates are more equal than others: Play fighting amongst adult subordinate male rats. *Aggressive Behavior, 19*(5), 385–393.

Pereira, A. G., Cruz, A., Lima, S. Q., & Moita, M. A. (2012). Silence resulting from the cessation of movement signals danger. *Current Biology, 22*(16), R627–R628.

Sales, G. D. (1991). The effect of 22 kHz calls and artificial 38 kHz signals on activity in rats. *Behavioural Processes, 24*(2), 83–93.

Shettleworth, S. (Ed.) (2010). *Cognition, evolution, and behavior* (pp. 466–506). Oxford/New York: Oxford University Press.

Thomas, D. A., Takahashi, L. K., & Barfield, R. J. (1983). Analysis of ultrasonic vocalizations emitted by intruders during aggressive encounters among rats (*Rattus norvegicus*). *Journal of Comparative Psychology, 97*(3), 201.

Wöhr, M., & Schwarting, R. K. (2007). Ultrasonic communication in rats: Can playback of 50-kHz calls induce approach behavior. *PLoS One, 2*(12), e1365.

Neuronal Correlates of Remote Fear Learning in Rats

Karolina Rokosz, Ewelina Knapska

Nencki Institute of Experimental Biology, Polish Academy of Sciences, Warsaw, Poland

EMOTIONAL CONTAGION—WHAT CAN ANIMALS LEARN FROM FEARFUL CONSPECIFICS?

Emotions are highly contagious. Emotional expressions of one individual can readily evoke similar emotions and behaviors in others, the phenomenon called *emotional contagion*. The synchronization of emotional expressions, vocalizations, postures, and movements is a very fast process of unconscious, automatic mimicry (Hatfield, Cacioppo, & Rapson, 1994). Such emotional convergence is commonly observed and evolutionarily conserved in the animal world. It has been suggested that adaptive significance of the ability to share emotions lies in the possibility of rapid adaptations to environmental challenges (Hatfield et al., 1994). Tuning one's emotional state to that of another, which increases the probability of similar behavior, is the form of social adaptation particularly important for emotions such as fear that signal a potential danger (Kelly, Iannone, & McCarty, 2016). Although one can learn about potentially harmful stimuli by directly experiencing an aversive event, interaction with a conspecific in danger provides information about threats in the environment without exposure to an immediate risk.

Mathematical models show that emotional contagion is often a more flexible strategy of coping with environmental threats than simple behavioral mimicry, that is, exact copying of the behavior of another individual (Nakahashi & Ohtsuki, 2015). Emotional contagion is particularly adaptive when the threat is not imminent yet; for example, when emotional expression of the demonstrator who had been attacked by a predator and escaped is warning others about the danger. The choice of strategy by the observer,

Neuronal Correlates of Empathy. http://dx.doi.org/10.1016/B978-0-12-805397-3.00009-7

be it fight, escape, freezing, or avoidance, requires evaluation of environmental conditions. To choose the most appropriate response, the observer has to carefully scan the environment to estimate the risk, which requires increased alertness and vigilance. Fear contagion clearly improves this ability.

Although emotional contagion is relatively well described at the behavioral level, much less is known about the neuronal mechanisms involved in this phenomenon. It has been proposed that the Mirror Neuron System, that is, the system active both when we feel a specific emotion, and when we see another person's emotional expressions, is a neural substrate responsible for emotional contagion (Keysers & Gazzola, 2009). However, this hypothesis is still lacking experimental support. To understand brain mechanisms underlying emotional contagion at the level of neurons, we need very simple animal models, which, unlike human or primate studies, allow manipulation of neuronal activity with great precision. In recent years, several models of emotional contagion have been described in rodents (Knapska et al., 2006; Meyza, Bartal, Monfils, Panksepp, & Knapska, 2016) providing an excellent starting point for studies on neurobiology of this phenomenon.

RODENT MODELS OF SOCIALLY TRANSFERRED FEAR

Rodents, particularly rats and mice, are commonly used in basic research as model animals. As these are social species, they are also regarded as well suited for studying interactions between conspecifics. Both rats and mice exhibit a wide range of social behaviors, including recognition of family (ingroup) members, agonistic behaviors required to defend and maintain territory and social rank, and maternal care (Barnett, 1958; The Mouse in Biomedical Research, 2006). Rats also display social buffering and prosocial behaviors (Ishii, Kiyokawa, Takeuchi, & Mori, 2016), as well as rescue behaviors (Ben-Ami Bartal, Decety, & Mason, 2011; Sato, Tan, Tate, & Okada, 2015).

Fear is an adaptive emotion that allows animals to respond appropriately to threats in the environment. It is evolutionarily conserved both at the behavioral and neuronal levels. The neural circuits that control fear expression and fear learning are relatively well understood (Maren, 2011). Thus most research aimed at understanding emotional contagion has focused on the capability to share fear. In recent years, several studies have shown that rats and mice are able to share fear socially (Panksepp & Lahvis, 2011; Panksepp & Panksepp, 2013). Various rodent models of fear contagion that differ in threat imminence have been developed. In models of imminent danger, the animal serving as a source of emotional stimulation (the demonstrator) is either subjected to classical fear conditioning or fear memory retrieval in close proximity of the observer animal. Both animals are in small enclosures, lacking the possibility to escape. In contrast, in remote danger models, the demonstrator, which had been subjected to fear conditioning, is transferred to a home territory where social interaction with an observer animal takes place. In this chapter, we focus on models of remote danger, while the models of imminent danger are reviewed in Chapter 8 (Jones and Monfils).

In contrast to protocols of imminent danger, in which both the demonstrator and the observer animals are put into an inescapable confinement and respond with passive fear responses, such as freezing and emission of distress calls, in the model of remote fear, the animals display active fear behaviors, such as exploration of the environment and risk assessment. It is noteworthy that, as the imminent and remote danger models seem to involve different coping strategies (passive and active, respectively), specific brain circuits that mediate these responses can differ (Bandler, Keay, Floyd, & Price, 2000; Keay & Bandler, 2001).

In an experimental *model of socially transferred fear* that we have designed (Knapska et al., 2006), rats are housed in pairs, and one member of the pair (the demonstrator) is removed and subjected to fear conditioning. After the fear-conditioning training, the emotionally aroused animal is returned to the home cage and allowed to interact with its naive cage mate (the observer, Fig. 9.1A). In the control group, the demonstrator is exposed to the experimental cage without fear conditioning. We showed that the demonstrator's arousal is socially transferred to the observer. The observers displayed both rapid increase in exploratory behavior and an increase in acoustic startle response, which is used as a measure of emotional arousal (Knapska et al., 2006). During the interaction, rats were engaged mainly in social exploratory behaviors, including sniffing and allogrooming of the body parts that excrete pheromones (Knapska, Mikosz, Werka, & Maren, 2010). They also produced significant amount of high-frequency vocalizations (>50 kHz, unpublished data).

The aforementioned protocol was recently adapted for studying emotional contagion in mice. As in rats, the observer mice displayed an increase in the number and duration of social contacts. Interestingly, BTBR T^+Itpr3^{tf}/J mice, a mouse model of idiopathic autism, displayed strikingly lower number and duration of social exploratory behaviors than the highly social C57BL/6J mice (Meyza et al., 2015).

The changes in the observer's behavior, that result from interaction with a fearful partner, show that to share another's fear, an individual does not have to be in direct danger, and that similar experience prior to the interaction is not required for the shared arousal to occur. Interaction with a recently fear conditioned conspecific triggered strong autonomic (increase in the heart rate, unpublished data) and behavioral responses (Knapska et al., 2006; Knapska et al., 2010), even in animals that had not been subjected to fear conditioning before. Thus it seems that interaction with a fearful partner allows *rapid* adaptation to environmental challenges. Another question is whether state matching in emotional contagion can exert *long-lasting effects* on behavior, affecting learning and memory.

(A)   (B)

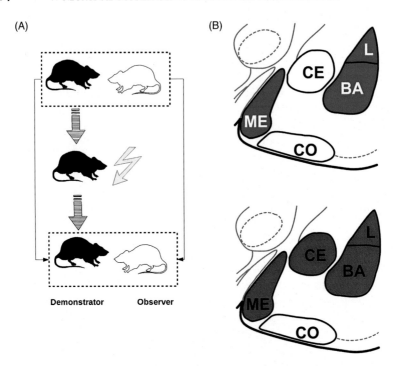

FIGURE 9.1   In the *model of socially transferred fear*, rats are housed in pairs (A). They are randomly assigned to S (Shocked) and NS (Non-Shocked) groups and, in each cage, one subject is chosen to be the demonstrator while the other becomes the observer. In the S group, the demonstrators are taken from their home cages and exposed to aversive conditioning (Pavlovian contextual fear conditioning). When the demonstrators are trained, their cohabitants (the observers) are kept in the home cages in a different, sound-attenuating room. Immediately after the training, the demonstrators are placed back in their home cages and allowed to interact with the observers. The NS (control) group is composed of rats treated similar to those from the S group, except that the NS demonstrators are exposed to the experimental cage without any training. The pattern of amygdala activation in the observers mirrors the one showed by the demonstrators except for the central amygdala, in which higher activation was seen in the observers (B).

SOCIAL MODULATION OF FEAR MEMORIES

The link among emotional expressions, empathic responses to these expressions, and collecting the knowledge about the environment remains unclear. However, there are several indications that socially transferred fear improves future performance of defensive responses exerting *long-lasting effects* on behavior. We found that a brief social interaction with a cage mate that has undergone an aversive learning experience promotes learning and memory of defensive responses in an otherwise naive rat (Knapska et al., 2010). Interaction with a fearful partner prior to the training session increased conditioned freezing measured on the following day in contextual fear conditioning task and facilitated both the acquisition and memory in a shock-motivated shuttle avoidance task. Learning enhancement observed in rats was not associated with a stress-induced increase in pain sensitivity or analgesia. In contrast, in mice, Bredy & Barad (2009) showed that exposure to a recently fear-conditioned animal impaired acquisition of cued conditioned fear. The reason for the discrepancy between the rats and mice experiments is not clear. It may stem from different levels of stress induced by interaction with a recently fear-conditioned partner or different coping strategies in these species. These hypotheses, however, require further studies.

In addition to affecting fear learning, social interaction with a recently fear-conditioned partner can also modulate extinction of conditioned fear (Bredy & Barad, 2009; Nowak, Werka, & Knapska, 2013). However, the results of studies carried to date do not offer a clear picture of this effect. While home cage interaction with a recently fear-conditioned partner facilitated extinction of conditioned fear response in mice, an exposure to recently fear-extinguished mouse (but not its urine alone) impaired it (Bredy & Barad, 2009). In contrast, an exposure (in an adjacent chamber of a shuttle box) during extinction memory retention test to a mouse showing high level of fear but not to a mouse showing low level of fear impaired fear extinction memory retrieval (Nowak et al., 2013). These discrepancies may result from different experimental protocols, in particular, from different handling procedures, leading to different levels of baseline stress during the experiments in the studies mentioned earlier.

In summary, emotional contagion not only provides information about present environmental threats, but also modulates learning of defensive responses and allows avoidance of dangerous situations in the future. However, more research is needed to understand the mechanisms through which social interaction improves or impairs learning.

NEURONAL CORRELATES OF SOCIALLY TRANSFERRED FEAR

Fear contagion, in contrast to more complex empathic behaviors, does not require higher-order emotional or cognitive control and can be observed also in species without well-developed cortical circuits. This suggests that we should look for the evolutionary origins of neuronal correlates of socially transferred fear in the subcortical parts of the limbic system, shared by all vertebrates. Thus we focused on the amygdala, the key structure in the brain that processes emotions. As the amygdala is a heterogeneous structure with several nuclei playing different roles in control of emotional behavior, we investigated their activation separately using c-Fos as a marker of neuronal activation. c-Fos is a protein transiently expressed in nuclei of activated neurons and thus offers a single-cell resolution (Knapska, Radwanska, Werka, & Kaczmarek, 2007). We compared activation patterns of different parts of the amygdala between rats that interacted with fearful partners (immediately after their aversive experience) and rats that interacted with partners merely exposed to a novel environment. We found that interaction with a fearful partner significantly increased activation of the lateral (LA), basal (BA), basomedial, and medial nuclei of the amygdala. The level of activation was comparable to that observed in rats directly exposed to footshocks during fear conditioning procedure (Fig. 9.1B). Interestingly, in the central nucleus (CeA), the activation was higher in the observers (Knapska et al., 2006; Mikosz, Nowak, Werka, & Knapska, 2015), suggesting that there is a group of neurons in this part of the amygdala that is activated by socially transferred but not by classical fear.

In mice, analysis of c-Fos expression revealed increased activation of the observers' amygdala, but it was limited to the BA. In contrast to rats, in the CeA, the number of c-Fos-positive nuclei was increased in the demonstrators, but not in the observers. This species specificity of the amygdala activation may reflect different strategies of fear transmission employed by rats and mice; however, further studies are needed to test this hypothesis. Interestingly, in the BTBR T (+) Itpr3(tf)/J mice, reduced social contact was associated with decreased activation of the BA, LA, and CeA (Meyza et al., 2015).

In rats, fear conditioning itself did not increase the number of c-Fos-positive neurons in the CeA, compared to the control rats, which were exposed to the cage without the training. This finding is in line with the results of previous studies (Knapska et al., 2007). In contrast, interaction with a recently fear-conditioned partner clearly increased expression of c-Fos in the CeA. Activation of additional subset of neurons can be associated with the active coping strategies displayed by the observers. This hypothesis is supported by the results, which showed that treatment with an anxiolytic (diazepam) before exposure to the fear-conditioned context reduced freezing, but enhanced active coping behaviors, as well as c-Fos expression in the CeA (Beck & Fibiger, 1995).

Central Amygdala

The central nucleus of the amygdala (CeA) is critical in fear learning (Ciocchi et al., 2010; Goosens & Maren, 2001; Haubensak et al., 2010; LeDoux, 2000) and controls both passive and active defensive responses (Gozzi et al., 2010). The CeA is the principal output structure of the amygdala. It is morphologically and functionally very heterogeneous and is often divided into the lateral (CeL) and medial (CeM) parts. The majority of output neurons that project to the brainstem and hypothalamic structures, mediating the endocrine, autonomic, and motor emotional responses, is located in the CeM (Cassell, Gray, & Kiss, 1986; Hopkins & Holstege, 1978; Veening, Swanson, & Sawchenko, 1984). However, a subpopulation of the CeL neurons projects directly to the output structures in the brain stem, in a pathway that could bypass the CeM and control emotional responses, including fear (Cassell et al., 1986; Gray & Magnuson, 1987; Koob, 2008). Our results show that both CeL and CeM in rats are activated by socially transferred fear (Mikosz et al., 2015).

The majority of CeA neurons are GABAergic (Cassell et al., 1986; McDonald, 1982) and express a variety of neuromodulators and their receptors (Stoop, Hegoburu, & van den Burg, 2015). Based on optogenetic, pharmacological, and electrophysiological studies, several subpopulations of neurons in the CeA have been identified as distinct neuronal microcircuits that control different types of defensive responses (e.g., active vs. passive fear responses), as well as reduce or increase fear, according to the environmental requirements (Gozzi et al., 2010; Haubensak et al., 2010).

Recently, we have used *c-fos*-driven targeting of channelrhodopsin into neurons involved in social interaction to specifically manipulate their activity during subsequent tests. The results showed that optogenetic activation of CeA neurons involved during interaction with recently fear-conditioned partner decreased social exploratory behaviors and ultrasonic communication. During the stimulation, rats spent more time actively exploring their environment (Knapska, 2015). This pattern of behavior resembles an active fear response (Gozzi et al., 2010) and suggests that the CeA neurons are not only involved in classically conditioned fear but also in socially induced fear.

This short overview of our research illustrates how the newly developed technologies for manipulation of neuronal activity, such as optogenetics, can be used to identify the function of neuronal circuits involved in emotional contagion. Such circuits can be further characterized in terms of their connectivity with other parts of the brain with the use of genetic and viral tools, as well as advanced imaging techniques. The detailed analysis of their connectivity and identification of the entire wiring network of neural connections in the brain activated by emotional contagion will shed light on how the brain computes information related to socially transferred emotions.

The data on neuronal correlates of fear contagion described so far were obtained in experiments involving male rats. As sex differences have been suggested in empathic behaviors, we also compared male and female rats in the model of socially transferred fear.

SEX DIFFERENCES IN SOCIALLY TRANSFERRED FEAR

Sex differences in empathy are widely studied in humans (Gruene, Roberts, Thomas, Ronzio, & Shansky, 2015; Rueckert & Naybar, 2008; Schulte-Rüther, Markowitsch, Shah, Fink, & Piefke, 2008). It has been shown that empathic abilities vary across the menstrual cycle in women (Derntl, Hack, Kryspin-Exner, & Habel, 2013; Derntl et al., 2008; Guapo et al., 2009). In contrast, the literature on the effects of cyclic hormonal changes on empathic behaviors in animal models is sparse. Female rodents have been used in several studies, but the effects of systematic sex and estrus cycle phase have not been examined (Atsak et al., 2011; Ben-Ami Bartal et al., 2011; Jones, Riha, Gore, & Monfils, 2014; Langford et al., 2006).

We used our rat model of socially transferred fear to compare the behavioral consequences and neural activation of emotional contagion in male and female rats (Mikosz et al., 2015). Though male and female observers behaved similarly during social interaction with a fear-conditioned demonstrator, being involved mainly in social exploratory behaviors directed at their partners, the interaction positively modulated two-way avoidance learning in male and diestral female rats, but not in estral females. In addition, the memory of avoidance response was worse in estral females. Consistent with our previous results, male rats paired with recently fear-conditioned partners had increased c-Fos expression in the CeA and LA. Moreover, we found increased activation of the prefrontal cortex. In contrast, in females no such changes were observed. Collectively, the results showed differences related to sex and estrus cycle phase in the susceptibility to fear contagion and underlying neuronal activation, suggesting the additional variables that should be taken into consideration in studies of socially transferred emotions.

SUMMARY

We observed that fear shared by conspecifics results in an increase of vigilance toward the environment and enhancement of subsequent fear learning. At the neuronal level, the results obtained so far show that the patterns of activation of amygdala in rats interacting with recently fear-conditioned partners generally paralleled that of rats directly exposed to aversive stimulation, suggesting an affective resonance between animals. It is, however, not clear, whether the activated neurons are the same for socially and nonsocially driven fear, as predicted by the Mirror Neuron System theory. Our results do not preclude such a possibility. However, finding a group of neurons in the CeA activated by social interaction with a fearful partner but not by classical fear conditioning shows that, mirroring is not the only mechanism involved. Certainly, to understand the mechanisms by which brain controls vicarious emotions and resonance behaviors, we need data obtained with single-cell resolution and techniques that allow investigating the function of identified neurons.

Simple animal models of fear contagion combined with advanced techniques of neurobiology provide a unique opportunity to fully appreciate the neuronal complexities of empathy, as well as to answer the question whether the same neuronal circuits are activated by social and nonsocial emotions. However, to interpret correctly the animal findings, much research in the field is needed. A key open question that future studies will need to consider is the translatability of the findings. Parallel studies on fear contagion are required to answer the question whether the systems for the social communication of emotions are homologous in humans and rodents and what is the evolutionary history of the phenomenon.

References

Atsak, P., Orre, M., Bakker, P., Cerliani, L., Roozendaal, B., Gazzola, V., et al. (2011). Experience modulates vicarious freezing in rats: A model for empathy. *PloS One, 6*(7), e21855.

Bandler, R., Keay, K. A., Floyd, N., & Price, J. (2000). Central circuits mediating patterned autonomic activity during active vs. passive emotional coping. *Brain Research Bulletin, 53*(1), 95–104.

Barnett, S. A. (1958). Physiological effects of social stress in wild rats. I. The adrenal cortex. *Journal of Psychosomatic Research, 3*(1), 1–11.

Beck, C. H., & Fibiger, H. C. (1995). Conditioned fear-induced changes in behavior and in the expression of the immediate early gene c-fos: With and without diazepam pretreatment. *The Journal of Neuroscience: The Official Journal of the Society for Neuroscience, 15*(1 Pt 2), 709–720.

Ben-Ami Bartal, I., Decety, J., & Mason, P. (2011). Empathy and pro-social behavior in rats. *Science (New York, N.Y.), 334*(6061), 1427–1430.

Bredy, T. W., & Barad, M. (2009). Social modulation of associative fear learning by pheromone communication. *Learning & Memory (Cold Spring Harbor, N.Y.), 16*(1), 12–18.

Cassell, M. D., Gray, T. S., & Kiss, J. Z. (1986). Neuronal architecture in the rat central nucleus of the amygdala: A cytological, hodological, and immunocytochemical study. *The Journal of Comparative Neurology, 246*(4), 478–499.

Ciocchi, S., Herry, C., Grenier, F., Wolff, S. B. E., Letzkus, J. J., Vlachos, I., et al. (2010). Encoding of conditioned fear in central amygdala inhibitory circuits. *Nature, 468*(7321), 277–282.

Derntl, B., Hack, R. L., Kryspin-Exner, I., & Habel, U. (2013). Association of menstrual cycle phase with the core components of empathy. *Hormones and Behavior, 63*(1), 97–104.

Derntl, B., Windischberger, C., Robinson, S., Lamplmayr, E., Kryspin-Exner, I., Gur, R. C., et al. (2008). Facial emotion recognition and amygdala activation are associated with menstrual cycle phase. *Psychoneuroendocrinology, 33*(8), 1031–1040.

Goosens, K. A., & Maren, S. (2001). Contextual and auditory fear conditioning are mediated by the lateral, basal, and central amygdaloid nuclei in rats. *Learning & Memory (Cold Spring Harbor, N.Y.), 8*(3), 148–155.

Gozzi, A., Jain, A., Giovannelli, A., Giovanelli, A., Bertollini, C., Crestan, V., et al. (2010). A neural switch for active and passive fear. *Neuron, 67*(4), 656–666.

Gray, T. S., & Magnuson, D. J. (1987). Neuropeptide neuronal efferents from the bed nucleus of the stria terminalis and central amygdaloid nucleus to the dorsal vagal complex in the rat. *The Journal of Comparative Neurology, 262*(3), 365–374.

Gruene, T. M., Roberts, E., Thomas, V., Ronzio, A., & Shansky, R. M. (2015). Sex-specific neu-roanatomical correlates of fear expression in prefrontal-amygdala circuits. *Biological Psychiatry, 78*(3), 186–193.

Guapo, V. G., Graeff, F. G., Zani, A. C. T., Labate, C. M., dos Reis, R. M., & Del-Ben, C. M. (2009). Effects of sex hormonal levels and phases of the menstrual cycle in the processing of emotional faces. *Psychoneuroendocrinology, 34*(7), 1087–1094.

Hatfield, E., Cacioppo, J. T., & Rapson, R. L. (1994). *Emotional contagion.* Cambridge University Press.

Haubensak, W., Kunwar, P. S., Cai, H., Ciocchi, S., Wall, N. R., Ponnusamy, R., et al. (2010). Genetic dissection of an amygdala microcircuit that gates conditioned fear. *Nature, 468*(7321), 270–276.

Hopkins, D. A., & Holstege, G. (1978). Amygdaloid projections to the mesencephalon, pons and medulla oblongata in the cat. *Experimental Brain Research, 32*(4), 529–547.

Ishii, A., Kiyokawa, Y., Takeuchi, Y., & Mori, Y. (2016). Social buffering ameliorates conditioned fear responses in female rats. *Hormones and Behavior, 81*, 53–58.

Jones, C. E., Riha, P. D., Gore, A. C., & Monfils, M. -H. (2014). Social transmission of Pavlovian fear: fear-conditioning by-proxy in related female rats. *Animal Cognition, 17*(3), 827–834.

Keay, K. A., & Bandler, R. (2001). Parallel circuits mediating distinct emotional coping reactions to different types of stress. *Neuroscience and Biobehavioral Reviews, 25*(7–8), 669–678.

Kelly, J. R., Iannone, N. E., & McCarty, M. K. (2016). Emotional contagion of anger is automatic: An evolutionary explanation. *The British Journal of Social Psychology/the British Psychological Society, 55*(1), 182–191.

Keysers, C., & Gazzola, V. (2009). Expanding the mirror: Vicarious activity for actions, emotions, and sensations. *Current Opinion in Neurobiology, 19*(6), 666–671.

Knapska E.,(2015). OASIS http://www.abstractsonline.com/plan/ViewAbstract. aspx?cKey=5e81fd69-0201-4276-b728-06901a863611&mID=3744&mKey=d0ff4555-8574-4fbb-b9d4-04eec8ba0c84&sKey=a7f7fea9-446c-4826-a5f9-e527aa1eff41.

Knapska, E., Mikosz, M., Werka, T., & Maren, S. (2010). Social modulation of learning in rats. *Learning & Memory (Cold Spring Harbor, N.Y.), 17*(1), 35–42.

Knapska, E., Nikolaev, E., Boguszewski, P., Walasek, G., Blaszczyk, J., Kaczmarek, L., et al. (2006). Between-subject transfer of emotional information evokes specific pattern of amygdala activation. *Proceedings of the National Academy of Sciences of the United States of America, 103*(10), 3858–3862.

Knapska, E., Radwanska, K., Werka, T., & Kaczmarek, L. (2007). Functional internal complexity of amygdala: Focus on gene activity mapping after behavioral training and drugs of abuse. *Physiological Reviews, 87*(4), 1113–1173.

Koob, G. F. (2008). A role for brain stress systems in addiction. *Neuron, 59*(1), 11–34.

Langford, D. J., Crager, S. E., Shehzad, Z., Smith, S. B., Sotocinal, S. G., Levenstadt, J. S., et al. (2006). Social modulation of pain as evidence for empathy in mice. *Science (New York, N. Y.), 312*(5782), 1967–1970.

LeDoux, J. E. (2000). Emotion circuits in the brain. *Annual Review of Neuroscience, 23*, 155–184.

Maren, S. (2011). Seeking a spotless mind: Extinction, deconsolidation, and erasure of fear memory. *Neuron, 70*(5), 830–845.

McDonald, A. J. (1982). Cytoarchitecture of the central amygdaloid nucleus of the rat. *The Journal of Comparative Neurology, 208*(4), 401–418.

Meyza, K., Nikolaev, T., Kondrakiewicz, K., Blanchard, D. C., Blanchard, R. J., & Knapska, E. (2015). Neuronal correlates of asocial behavior in a BTBR T (+) Itpr3(tf)/J mouse model of autism. *Frontiers in Behavioral Neuroscience, 9*, 199.

Meyza, K. Z., Bartal, I. B. -A., Monfils, M. H., Panksepp, J. B., & Knapska, E. (2016). The roots of empathy: Through the lens of rodent models. *Neuroscience and Biobehavioral Reviews.*

Mikosz, M., Nowak, A., Werka, T., & Knapska, E. (2015). Sex differences in social modulation of learning in rats. *Scientific Reports, 5*, 18114.

Nakahashi, W., & Ohtsuki, H. (2015). When is emotional contagion adaptive? *Journal of Theoretical Biology, 380*, 480–488.

Nowak, A., Werka, T., & Knapska, E. (2013). Social modulation in extinction of aversive memories. *Behavioural Brain Research, 238,* 200–205.

Panksepp, J. B., & Lahvis, G. P. (2011). Rodent empathy and affective neuroscience. *Neuroscience and Biobehavioral Reviews, 35*(9), 1864–1875.

Panksepp, J., & Panksepp, J. B. (2013). Toward a cross-species understanding of empathy. *Trends in Neurosciences, 36*(8), 489–496.

Rueckert, L., & Naybar, N. (2008). Gender differences in empathy: The role of the right hemisphere. *Brain and Cognition, 67*(2), 162–167.

Sato, N., Tan, L., Tate, K., & Okada, M. (2015). Rats demonstrate helping behavior toward a soaked conspecific. *Animal Cognition, 18*(5), 1039–1047.

Schulte-Rüther, M., Markowitsch, H. J., Shah, N. J., Fink, G. R., & Piefke, M. (2008). Gender differences in brain networks supporting empathy. *NeuroImage, 42*(1), 393–403.

Stoop, R., Hegoburu, C., & van den Burg, E. (2015). New opportunities in vasopressin and oxytocin research: a perspective from the amygdala. *Annual Review of Neuroscience, 38,* 369–388.

The mouse in biomedical research: history, wild mice, and genetics. (2006). Academic Press.

Veening, J. G., Swanson, L. W., & Sawchenko, P. E. (1984). The organization of projections from the central nucleus of the amygdala to brainstem sites involved in central autonomic regulation: A combined retrograde transport-immunohistochemical study. *Brain Research, 303*(2), 337–357.

10

Lost in Translation: Improving Our Understanding of Pain Empathy

Sivaani Sivaselvachandran,
Meruba Sivaselvachandran, Salsabil Abdallah,
Loren J. Martin

University of Toronto Mississauga, Mississauga, ON, Canada

INTRODUCTION

Given the complexity of empathy, investigating its neurobiological underpinnings would be meaningless without breaking it down into component processes. An increasing amount of evidence suggests that even the most advanced forms of empathy in humans are built upon more basic forms (Preston & De Waal, 2002). From a basic science perspective, this becomes really interesting and suggests that using animal models to understand the core components of empathy may help us develop a richer and deeper understanding of human empathy. As such, interest in the topic of animal empathy has grown tremendously over the last decade with data suggesting that the central features of empathy are shared among rodents and nonhuman mammals (Martin, Tuttle, & Mogil, 2014; Sivaselvachandran, Acland, Abdallah, & Martin, 2016).

In studies of empathy, observational pain, which is observing or witnessing another in pain, is often used because pain elicits a set of distinct and observable behaviors that are easily understood and measured. The act of witnessing another in pain activates the same brain regions as first-hand pain experiences (Zaki, Wager, Singer, Keysers, & Gazzola, 2016). Neural activity comparably increases within the anterior insula and the cingulate cortex during both the sensory and observational experiences of

Neuronal Correlates of Empathy. http://dx.doi.org/10.1016/B978-0-12-805397-3.00010-3

pain (Singer et al., 2004). Observing another in pain has also been shown to activate parts of the somatosensory and motor cortices, brain regions that are critical for pain detection and initiating the appropriate withdrawal responses (Avenanti, Bueti, Galati, & Aglioti, 2005). Furthermore, observing others in pain can intensify pain perception in both mice and people (Martin, Hathaway, et al., 2015), while common analgesics such as acetaminophen reduce the ability to relate to both the physical and social pains experienced by others (Mischkowski, Crocker, & Way, 2016). The present review briefly discusses the social modulation of pain and the contribution of the affective pain system to the experience of empathy. It then considers the translatability of findings within this realm and approaches to study empathy that would benefit from rodent → human or human → rodent investigations.

THE SOCIAL NEUROSCIENCE OF PAIN IN RODENTS

Few studies in the social neurosciences have focused on social stimuli that directly modulate pain. In the laboratory interpersonal factors, such as the presence, behavior and spatial proximity of an observer are known to affect pain sensitivity. Using the key search terms, such as "pain," "interpersonal," "empathy," "attachment," "social context," "social interaction," "social support," "social presence," and "social modulation," Krahe, Springer, Weinman, and Fotopoulou (2013) reported on only 26 studies that directly examined social factors that modulate pain sensitivity in the laboratory. The emergent results are rather complex given the variability in experimental framework and procedures, including vast differences among sample populations, social partners, pain assays, and pain measures. However, the results point toward important interacting factors such as catastrophizing, attachment style, pain syndrome, and the context of interaction as determining outcomes. There is also evidence that people develop pain-related beliefs and behaviors through observational learning (Goubert, Vlaeyen, Crombez, & Craig, 2011), a fundamental social phenomenon that is preserved in many animal species (Dawson, Avargues-Weber, Chittka, & Leadbeater, 2013; Isbaine, Demolliens, Belmalih, Brovelli, & Boussaoud, 2015; Jeon et al., 2010).

Evidence of empathy-like behaviors in different animal species is widespread and range from sympathy in elephants to helping behavior in dolphins. These findings mostly stem from anecdotal observations, but researchers are becoming increasingly interested in the reliable measurement and interpretation of such behaviors. Scientists are now finding more and more clever ways to study these behaviors in controlled laboratory environments (Ben-Ami Bartal, Decety, & Mason, 2011; Burkett et al., 2016; Langford et al., 2006; Sato, Tan, Tate, & Okada, 2015) and specifically the use of rodents has started to allow for the investigation of empathy on a level that has never before been possible (Chen, Panksepp, & Lahvis, 2009; Jeon et al., 2010; Martin, Hathaway, et al., 2015). Most rodent studies can be divided into two main categories: emotional contagion (social modulation of fear and pain) and prosociality (helping others in distress) (see Sivaselvachandran et al., 2016 for a review). In rodents, we simply do not know whether these behaviors are representative of human empathy, so we label them as emotional contagion or empathy-like behaviors. Emotional contagion is a fundamental capacity that enables the sharing of emotional information and has come to dominate the animal empathy jargon as most researchers agree that animals possess this capability. However, emotional contagion is not the only empathy-like capability exhibited by rodents and behaviors such as social buffering, social priming, social transfer, and social analgesia/hyperalgesia also fall within the realm of empathy-like phenomena (Martin et al., 2014; Panksepp & Lahvis, 2011; Sivaselvachandran et al., 2016).

The empirical evidence unquestionably suggests that rodents are capable of low-level forms of empathic responding, but much remains to be unraveled in terms of sensory modalities, neuroanatomy, neurophysiology, and of course social relationships. The evidence collected to date strongly suggests that systems that modulate empathy-like behaviors in rodents are the same as those that seem to be involved in humans. Formative support reinforces the importance of brain regions such as the paraventricular nucleus of the hypothalamus (Sanna, Argiolas, & Melis, 2012), amygdala (Knapska et al., 2006), and the medial prefrontal cortex (Li et al., 2014) as critical to not only rodent empathy, but also human empathy (see Singer & Lamm, 2009 for a review). The real advantage of using rodents, however, is that more precise mechanisms can be worked out. For instance, the field has begun to connect empathy behavior in rodents to $Ca_v1.2$ calcium channels in the anterior cingulate cortex (Jeon et al., 2010), $GABA_A$ receptors in the locus coeruleus (Bravo et al., 2013), as well as serotonin-1A receptors (Ago, Takuma, & Matsuda, 2014), and μ-opioid receptors (D'Amato & Pavone, 2012). Although directly comparing neural activity and behavioral patterns between human and animal studies are not necessary, this just makes for better science and arguably a more complete picture.

Studies that explore social transference of fear in rodents explore a core feature of empathy, that is, the ability to share emotional experiences (Chen et al., 2009; Jeon et al., 2010; Jeon & Shin, 2011; Kavaliers, Choleris, & Colwell, 2001). In one example of this phenomenon, a mouse is placed in a cage with biting flies, while another mouse observes this interaction (Kavaliers, Choleris, et al, 2001; Kavaliers, Colwell, & Choleris, 2001). The next day observer mice are exposed to biting flies whose biting appendages have been removed. Observer mice showed distress-like behavioral patterns similar to those exhibited by mice that were actually bitten by the flies (i.e., avoidance, analgesia). This is a classic display of emotional contagion, as the paradigm requires one mouse to recognize the distress of another, this distress is then shared/transferred to a naive mouse resulting in similar behavioral patterns. Jeon et al. (2010) also studied the emotional transference of fear expression in mice by using a painful footshock as the conditioning stimulus. In their paradigm (which has been used by many laboratories around the world), one mouse receives repetitive footshocks, while another mouse observes this process (usually through a transparent barrier). Typically, observer mice are found to have higher fear responses when observing their mating partners or siblings receive footshocks compared with unfamiliar mice. This is indicative of empathy, as the social transfer of fear in rodents was stronger among familiars, falling in line with the principles of emotional contagion and state sharing.

In rodents, much like people, the observation of another in pain affects how the observer responds to the pain of the other. For example, a female, but not a male mouse will approach another mouse exhibiting obvious signs of pain and discomfort, which leads to reduced pain behaviors from the mouse in pain (Langford, Tuttle, et al., 2010). The heightened response to social approach in females could be related to an amplified affiliation mechanism. The "tend and befriend" model proposed by Taylor et al. (2000) describes that under conditions of stress, female, but not male mammals have a tendency to care for their young and engage in social interactions. Exposure to a distressed mouse can be viewed as mild stress for an observer mouse, which increases social approach, or in other words empathy-like behaviors, toward the distressed conspecific. As with the majority of studies on empathy, familiarity seems to be key as beneficial effects of social contact were seen only when the approaching mouse was a cagemate of the mouse in pain rather than a stranger (Langford, Tuttle, et al., 2010). Interestingly, pain behaviors in mice are enhanced by the observation of another mouse in pain and this is also dependent on social familiarity (Langford et al., 2006; Li et al., 2014). However, threatening social encounters from unfamiliar male mice have been shown to reduce pain sensitivity in a testosterone-dependent fashion (Langford et al., 2011) and the social reunion of male siblings has been shown to be analgesic, a reaction dependent on the endogenous opioid system (D'Amato & Pavone, 2012).

In rodents, the sensory modalities that communicate pain are still not clear although a combination of visual and olfactory cues are purportedly responsible (Langford et al., 2006). That pain status is communicated visually in rodents is intriguing as rodents have poorly developed visual abilities and are considered "more olfactory" in nature (Brown & Wong, 2007). However, a rodent in pain displays clearly visible pain behaviors, including changes in body position, licking afflicted areas, and altered facial expressions (Langford, Bailey, et al., 2010; Martin, Hathaway, et al., 2015;

Martin, Piltonen, et al., 2015). In fact, male rats display an aversion to the image of another rat in pain, and a combination of both the face and the body are required for pain to be properly communicated (Nakashima, Ukezono, Nishida, Sudo, & Takano, 2015). In addition, whether a rodent approaches or avoids another rodent in pain may be entirely related to the sex of the animal because approach to pain was only observed in females (Langford, Tuttle, et al., 2010) and avoiding painful rat faces was tested using only male subjects (Nakashima et al., 2015).

Perhaps approach to pain, especially if it subserves some benefit (i.e., analgesia), might qualify as helping or altruistic behavior because presumably pain is being communicated for the sole purpose of signaling danger and thus pain avoidance would seem to be the more appropriate response. However, it has been shown that rats have the ability to learn basic operant procedures to help distressed companions. In most of these circumstances, it is the stress of the distressed individual that compels rats to engage in prosocial behaviors, which include opening restrainers to free trapped rats (Ben-Ami Bartal et al., 2011), pressing a lever to lower a squealing conspecific (Rice & Gainer, 1962), or opening a door that allows a water-soaked rat to escape (Sato et al., 2015). As it turns out, female rats are more likely to free cagemates compared with male rats, a feature that we have only touched upon, but is prevalent among most empathy-related behaviors. With increasing evidence of empathy in rodents and knowledge of pain pathways, there is great possibility of bridging the gap between empathy behavior and its underlying neural mechanisms.

TOWARD A TRANSLATIONAL NEUROSCIENCE OF EMPATHY

The capacity to communicate pain to others and to experience the pain of others by observation is fundamental to social interactions and relationships. Feelings of emotional pain can be directly transferred and often rely on the affective state of the individual (Jackson, Meltzoff, & Decety, 2005). There is increasing evidence that social relationships can be "painful" (Eisenberger & Lieberman, 2004; Eisenberger, Lieberman, & Williams, 2003; Iannetti & Mouraux, 2011) and somatic symptoms have been observed in people suffering from social pain (Zisook, Devaul, & Click, 1982) including whole body inflammation and increased activity within the dorsal anterior cingulate cortex and anterior insula (Eisenberger, Inagaki, Rameson, Mashal, & Irwin, 2009).

In humans, at a single-cell level, pain-related neurons in the anterior cingulate cortex exhibit similar discharge patterns during primary as well as observational pain (Hutchison, Davis, Lozano, Tasker, & Dostrovsky, 1999). In addition, observing the administration of noxious mechanical stimuli to another person evokes a distinct activity pattern in the right dorsal anterior cingulate cortex that is similar to primary pain (Morrison, Lloyd, di Pellegrino, & Roberts, 2004). In fact, the mere expectation of an impending painful stimulus to a social partner produces significant hemodynamic responses in the anterior insula, anterior cingulate cortex, brain stem, and cerebellum (Singer et al., 2004). In almost every study on social and observational pain in people, the anterior cingulate cortex and insula are activated or involved to a certain extent supporting an important role of these structures in perceiving and detecting another's pain.

The same also seems to be true for the animal models that are used to measure empathy and related behaviors. In these models, empathy is often associated with the direct transfer of emotional information, including fear, stress, and pain. In line with this, Jeon et al. (2010) have shown that the anterior cingulate cortex and brain structures that comprise the affective, but not the sensory pain system are necessary for observational fear learning. In prairie voles, fear/stress-mediated allogrooming is also dependent upon the anterior cingulate cortex and neurohormones such as oxytocin mediate this effect (Burkett et al., 2016). Furthermore, extensive limbic activation, primarily within the amygdala, has been reported following remote fear learning (Knapska et al., 2006). In rodent models of empathy, footshock-induced fear learning is the most widely used and accepted paradigm, and although a footshock is painful for a rodent, these models do not measure the direct transfer of painful information. This is unfortunate because the majority of human empathy studies use pain as both an independent and a dependent measure, whereas the majority of rodent studies measure fear as a proxy for empathy (Figs. 10.1 and 10.2). Of the human studies that specifically examined pain empathy, the majority measured the perception of another's pain (mostly using videos and pictures), while only a subset specifically assessed pain contagion in rodents (Fig. 10.2).

There seems to be a fundamental disconnect between clinical researchers and basic scientists. That is, empathy in people is predominately studied using pain as the preceding stimulus and dependent measure, while rodent studies overwhelmingly use fear-based approaches. Why is this? It most likely stems from the reliability, familiarity, and ease of using fear-conditioning systems, as they are believed to be more robust and reliable than injections of chemical irritants (i.e., acetic acid, formalin, bee venom, etc.). We would argue that pain should be used as both the primary stimulus *and* dependent measure so that studies on rodent empathy can capture pain variability and become more comparable to their human counterparts. We are not arguing that social fear learning is not a valid approach; it just may not be the best model to study empathy in light of the approaches used by clinical researchers. It is also likely that the neural circuits used to

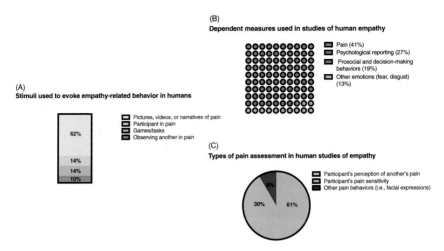

FIGURE 10.1   The variables and measures used to quantify empathy in laboratory studies using humans. We searched Pubmed for articles that contained either empathy, contagion, or prosocial within the title or abstract. This returned over 1400 unique articles. Of
these, we scanned the title and abstracts to include papers that examined empathy, contagion, or prosocial behaviors in a laboratory setting using human subjects. This resulted in a
final list of 156 relevant papers published between 1985 and 2016. (A) The relative frequency
of independent variables that are typically used to evoke empathy-related behaviors in laboratory studies of empathy. The majority of studies used some form of media as the primary
empathy manipulation (i.e., pictures, videos, narratives), while a participant's pain sensitivity was not frequently used. In addition, the studies that intended to measure the perception
of another person's pain only used another subject (i.e., not a video or picture) 10% of the
time. (B) The ratio of primary dependent measures for human empathy studies is presented.
Pain is the predominant dependent variable accounting for 41% of all studies; this includes
pain ratings, sensitivity, behaviors, and perception of another's pain. Psychological reporting (i.e., interpersonal reactivity index, implicit association task, etc.) accounted for 27% of
the primary measures, prosocial, and decision-making behaviors primarily from games/
tasks was used in 19% of studies, while other measures of emotions such as fear and disgust
accounted for only 13%. (C) The breakdown of the types of pain assessment is presented. The
majority of studies used the perception of another's pain as the primary measure, while only
30% actually measured changes in pain sensitivity of the subject. In addition, a small propor-
tion of studies quantify additional pain behaviors such as facial expressions and touching
affected body parts.

decode empathy for heightened pain sensitivity vastly differ from those
used to encode empathy from other emotions such as fear. For instance,
the cortical activation patterns that are observed in pain empathy studies
(i.e., anterior cingulate and insular cortices) may not reflect the specific
neural correlates of empathy "for fear" per se, but neural activation relating to more general feelings of compassion toward another in distress. Accordingly, exposure to others' emotional distress would lead to the sensi-
tization of areas that are involved in the processing of noxious stimuli and
this may, in turn, apply to fear- and stress-related stimuli. It is important to

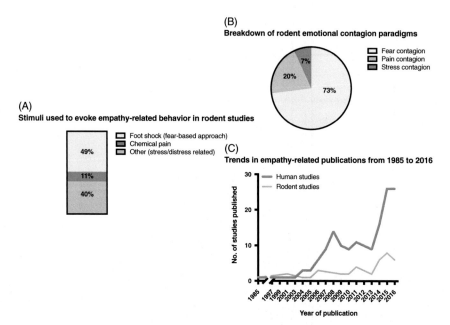

FIGURE 10.2   The variables and measures used to quantify empathy in laboratory studies using rodents. We scanned the title and abstract of the papers from our initial search (Fig. 10.1) to include papers that examined empathy, contagion, or prosocial behaviors using rodent models. This resulted in a final list of 43 relevant papers published between 1985 and 2016. (A) Approximately half of the studies that used rodent models to study empathy-related behaviors use a fear-based approach (49%), while only 11% used a chemical pain stimulus. Other stimuli accounted for approximately 40% often using restraint stress, predator-related stress, or water stress to evoke empathy-related behaviors. (B) Breaking down emotional contagion into different types of emotions revealed that fear contagion is the most popular approach followed by pain and then stress. (C) Interest in the field of empathy as it relates to humans and rodents is growing. From the years 1985 to 2016, an upward trend toward more empathy-related studies in humans and rodents is shown. Although both fields have witnessed a surge within the past few years, human studies greatly outnumber rodent studies in this domain.

bear in mind that we may all be measuring the same thing or we may not be. Beyond cortical activations, we simply do not know whether empathy for pain is the same as empathy for fear or related emotions.

In 2006, Langford et al. arguably started the resurgence and interest into using rodents for the study of empathy-related behaviors. Mice in pain, tested in the presence of cagemates—also in pain—displayed more pain behavior than mice tested alone or in the presence of strangers. We have published subsequent work demonstrating that these findings translate in a surprisingly direct manner to humans, that is, subjects report higher pain levels when tested in the presence of a friend in pain, but not a stranger (Martin, Hathaway, et al., 2015). We have also started to work out the neurohormones involved and have found that stress blocks the capacity for emotional contagion of pain in mouse and human strangers. In humans, empathy itself increases pain perception (Loggia, Mogil, & Bushnell, 2008), as does the presence of a solicitous spouse (Flor, Kerns, & Turk, 1987), suggesting that measuring pain behaviors in rodents should be given more consideration when designing studies of rodent empathy. Science in general would benefit from translational studies that focus on either rodent → human or human → rodent investigations. These types of investigations will open the doors that allow for the use of mouse genetics and molecular techniques to better understand the mechanisms on which more advanced social behaviors are built. Much work remains to be done, but the underlying concepts have larger implications than empathy or emotional transference. Explicit investigation of pain as it relates to the social neurosciences in general may lead to more fruitful and successful treatments of chronic pain in humans. The parallels that exist between humans and rodents in the social neuroscience of pain and empathy indicate that translational research is possible within this domain and should not be dismissed as either irrelevant or artefactual.

CONCLUSION

Pain is universal. It has no bounds. There is no race, culture, or sex that is immune to pain. Pain is the one modality that is equally a sensation and an emotion. In fact, it is the only sensory modality that does not habituate, but sensitizes over time with "pain-responsive" neurons becoming better at detecting and transmitting pain-related signals. Both the sensory and emotional aspects of pain are not solely determined by the degree of the painful stimulus, but rather "higher" cognitive activities that modulate perceived intensity and unpleasantness (Bushnell, Ceko, & Low, 2013). The social environment perhaps best exemplifies the existence of this higher order modulation, as there is increasing evidence that the presence, behavior and spatial proximity of an observer modulates pain (Krahe et al., 2013). Moreover, the emotional qualities of pain can be recapitulated in "pain-free" individuals by simply observing another in pain. When this happens, we ascribe these behaviors to empathy, sympathy, or some form of compassion.

At its core, empathy is a fundamental process required for social behavior. Species other than humans exhibit the basic characteristics of empathy, and we believe that in order to fully understand the neural processing behind this phenomenon, experiments conducted using both human and nonhuman species must be considered. The current review has specifically focused on traditional laboratory models (i.e., mice and rats) to understand empathy in humans and the converse approach where pain empathy in humans is driving rodent research. Ultimately, scientists that focus on pain empathy in humans or those studying emotional-state sharing in rodents need to work in a collaborative fashion to answer one of the most challenging questions in the neurosciences: What gives rise to empathy? Reconciling translational findings from rodent to humans or humans to rodents is challenging under the best of circumstances and designing experiments where multiple species are used is not always possible, but with the use of techniques such as fMRI and electrophysiological recordings, translational studies are more possible now than ever.

For us, the parallels that exist between humans and rodents are evident and should be exploited to advance the field. Rodents are capable of simple forms of the same types of social interactions that affect acute and chronic pain; similar social sex differences have been found in both species; such phenomena are dependent on familiarity between social actors and the reduction of social stress elicits emotional contagion in mouse and human strangers (Martin, Hathaway, et al., 2015; Mogil, 2012; Mogil, Davis, & Derbyshire, 2010). The social neuroscience of pain has the potential to make huge inroads into understanding complex disease states such as autism, psychopathy, and chronic pain treatment. That rodents are appropriate subjects for scientific study is an understatement; they have advanced social structures and behaviors and all of the machinery necessary to engage in sharing, caring, and compassion. Much research in this field is left to be explored, but most importantly the development of new models and paradigms will eventually help us to understand these complex and interesting behaviors.

References

Ago, Y., Takuma, K., & Matsuda, T. (2014). Potential role of serotonin1A receptors in post-weaning social isolation-induced abnormal behaviors in rodents. *Journal of Pharmaceutical Sciences, 125,* 237–241.
Avenanti, A., Bueti, D., Galati, G., & Aglioti, S. M. (2005). Transcranial magnetic stimulation highlights the sensorimotor side of empathy for pain. *Nature Neuroscience, 8,* 955–960.
Ben-Ami Bartal, I., Decety, J., & Mason, P. (2011). Empathy and pro-social behavior in rats. *Science, 334,* 1427–1430.
Bravo, L., Alba-Delgado, C., Torres-Sanchez, S., Mico, J. A., Neto, F. L., & Berrocoso, E. (2013). Social stress exacerbates the aversion to painful experiences in rats exposed to chronic pain: the role of the locus coeruleus. *Pain, 154,* 2014–2023.
Brown, R. E., & Wong, A. A. (2007). The influence of visual ability on learning and memory performance in 13 strains of mice. *Learning & Memory, 14,* 134–144.
Burkett, J. P., Andari, E., Johnson, Z. V., Curry, D. C., De Waal, F. B., & Young, L. J. (2016). Oxytocin-dependent consolation behavior in rodents. *Science, 351,* 375–378.
Bushnell, M. C., Ceko, M., & Low, L. A. (2013). Cognitive and emotional control of pain and its disruption in chronic pain. *Nature Reviews Neuroscience, 14,* 502–511.
Chen, Q., Panksepp, J. B., & Lahvis, G. P. (2009). Empathy is moderated by genetic background in mice. *PLoS One, 4,* e4387.

D'Amato, F. R., & Pavone, F. (2012). Modulation of nociception by social factors in rodents: contribution of the opioid system. *Psychopharmacology (Berl)*, *224*, 189–200.

Dawson, E. H., Avargues-Weber, A., Chittka, L., & Leadbeater, E. (2013). Learning by observation emerges from simple associations in an insect model. *Current Biology*, *23*, 727–730.

Eisenberger, N. I., Inagaki, T. K., Rameson, L. T., Mashal, N. M., & Irwin, M. R. (2009). An fMRI study of cytokine-induced depressed mood and social pain: the role of sex differences. *Neuroimage*, *47*, 881–890.

Eisenberger, N. I., & Lieberman, M. D. (2004). Why rejection hurts: a common neural alarm system for physical and social pain. *Trends in Cognitive Sciences*, *8*, 294–300.

Eisenberger, N. I., Lieberman, M. D., & Williams, K. D. (2003). Does rejection hurt? An FMRI study of social exclusion. *Science*, *302*, 290–292.

Flor, H., Kerns, R. D., & Turk, D. C. (1987). The role of spouse reinforcement, perceived pain, and activity levels of chronic pain patients. *Journal of Psychosomatic Research*, *31*, 251–259.

Goubert, L., Vlaeyen, J. W., Crombez, G., & Craig, K. D. (2011). Learning about pain from others: an observational learning account. *Journal of Pain*, *12*, 167–174.

Hutchison, W. D., Davis, K. D., Lozano, A. M., Tasker, R. R., & Dostrovsky, J. O. (1999). Pain-related neurons in the human cingulate cortex. *Nature Neuroscience*, *2*, 403–405.

Iannetti, G. D., & Mouraux, A. (2011). Can the functional MRI responses to physical pain really tell us why social rejection "hurts"? *Proceedings of the National Academy of Sciences*, *108*, E344.

Isbaine, F., Demolliens, M., Belmalih, A., Brovelli, A., & Boussaoud, D. (2015). Learning by observation in the macaque monkey under high experimental constraints. *Behavioural Brain Research*, *289*, 141–148.

Jackson, P. L., Meltzoff, A. N., & Decety, J. (2005). How do we perceive the pain of others? A window into the neural processes involved in empathy. *Neuroimage*, *24*, 771–779.

Jeon, D., Kim, S., Chetana, M., Jo, D., Ruley, H. E., Lin, S. Y., Rabah, D., Kinet, J. P., & Shin, H. S. (2010). Observational fear learning involves affective pain system and Cav1. 2 Ca2+ channels in ACC. *Nature Neuroscience*, *13*, 482–488.

Jeon, D., & Shin, H. S. (2011). A mouse model for observational fear learning and the empathetic response. *Current Protocols in Neuroscience*, *57*, 8.27.1–8.27.9 Chapter 8.

Kavaliers, M., Choleris, E., & Colwell, D. D. (2001a). Learning from others to cope with biting flies: social learning of fear-induced conditioned analgesia and active avoidance. *Behavioral Neuroscience*, *115*, 661–674.

Kavaliers, M., Colwell, D. D., & Choleris, E. (2001b). NMDA-mediated social learning of fear-induced conditioned analgesia to biting flies. *Neuroreport*, *12*, 663–667.

Knapska, E., Nikolaev, E., Boguszewski, P., Walasek, G., Blaszczyk, J., Kaczmarek, L., & Werka, T. (2006). Between-subject transfer of emotional information evokes specific pattern of amygdala activation. *Proceedings of the National Academy of Sciences*, *103*, 3858–3862.

Krahe, C., Springer, A., Weinman, J. A., & Fotopoulou, A. (2013). The social modulation of pain: others as predictive signals of salience—a systematic review. *Frontiers in Human Neuroscience*, *7*, 386.

Langford, D. J., Bailey, A. L., Chanda, M. L., Clarke, S. E., Drummond, T. E., Echols, S., Glick, S., Ingrao, J., Klassen-Ross, T., Lacroix-Fralish, M. L., Matsumiya, L., Sorge, R. E., Sotocinal, S. G., Tabaka, J. M., Wong, D., van den Maagdenberg, A. M., Ferrari, M. D., Craig, K. D., & Mogil, J. S. (2010a). Coding of facial expressions of pain in the laboratory mouse. *Nature Methods*, *7*, 447–449.

Langford, D. J., Crager, S. E., Shehzad, Z., Smith, S. B., Sotocinal, S. G., Levenstadt, J. S., Chanda, M. L., Levitin, D. J., & Mogil, J. S. (2006). Social modulation of pain as evidence for empathy in mice. *Science*, *312*, 1967–1970.

Langford, D. J., Tuttle, A. H., Briscoe, C., Harvey-Lewis, C., Baran, I., Gleeson, P., Fischer, D. B., Buonora, M., Sternberg, W. F., & Mogil, J. S. (2011). Varying perceived social threat modulates pain behavior in male mice. *Journal of Pain*, *12*, 125–132.

Langford, D. J., Tuttle, A. H., Brown, K., Deschenes, S., Fischer, D. B., Mutso, A., Root, K. C., Sotocinal, S. G., Stern, M. A., Mogil, J. S., & Sternberg, W. F. (2010b). Social approach to pain in laboratory mice. *Social Neuroscience, 5,* 163–170.

Li, Z., Lu, Y. F., Li, C. L., Wang, Y., Sun, W., He, T., Chen, X. F., Wang, X. L., & Chen, J. (2014). Social interaction with a cagemate in pain facilitates subsequent spinal nociception via activation of the medial prefrontal cortex in rats. *Pain, 155,* 1253–1261.

Loggia, M. L., Mogil, J. S., & Bushnell, M. C. (2008). Empathy hurts: compassion for another increases both sensory and affective components of pain perception. *Pain, 136,* 168–176.

Martin, L. J., Hathaway, G., Isbester, K., Mirali, S., Acland, E. L., Niederstrasser, N., Slepian, P. M., Trost, Z., Bartz, J. A., Sapolsky, R. M., Sternberg, W. F., Levitin, D. J., & Mogil, J. S. (2015a). Reducing social stress elicits emotional contagion of pain in mouse and human strangers. *Current Biology, 25,* 326–332.

Martin, L. J., Piltonen, M. H., Gauthier, J., Convertino, M., Acland, E. L., Dokholyan, N. V., Mogil, J. S., Diatchenko, L., & Maixner, W. (2015b). Differences in the antinociceptive effects and binding properties of propranolol and bupranolol enantiomers. *Journal of Pain, 16,* 1321–1333.

Martin, L. J., Tuttle, A. H., & Mogil, J. S. (2014). The interaction between pain and social behavior in humans and rodents. *Current Topics in Behavioral Neurosciences, 20,* 233–250.

Mischkowski, D., Crocker, J., & Way, B. M. (2016). From painkiller to empathy killer: acetaminophen (paracetamol) reduces empathy for pain. *Social Cognitive and Affective Neuroscience, 11,* 1345–1353.

Mogil, J. S. (2012). The surprising empathic abilities of rodents. *Trends in Cognitive Sciences, 16,* 143–144.

Mogil, J. S., Davis, K. D., & Derbyshire, S. W. (2010). The necessity of animal models in pain research. *Pain, 151,* 12–17.

Morrison, I., Lloyd, D., di Pellegrino, G., & Roberts, N. (2004). Vicarious responses to pain in anterior cingulate cortex: is empathy a multisensory issue? *Cognitive, Affective, & Behavioral Neuroscience, 4,* 270–278.

Nakashima, S. F., Ukezono, M., Nishida, H., Sudo, R., & Takano, Y. (2015). Receiving of emotional signal of pain from conspecifics in laboratory rats. *Royal Society Open Science, 2,* 140381.

Panksepp, J. B., & Lahvis, G. P. (2011). Rodent empathy and affective neuroscience. *Neuroscience & Biobehavioral Reviews, 35,* 1864–1875.

Preston, S. D., & De Waal, F. B. (2002). Empathy: its ultimate and proximate bases. *Behavioral and Brain Sciences, 25*(1–20), 20–71.

Rice, G. E., & Gainer, P. (1962). Altruism in Albino Rat. *Journal of Comparative and Physiological Psychology, 55,* 123.

Sanna, F., Argiolas, A., & Melis, M. R. (2012). Oxytocin-induced yawning: sites of action in the brain and interaction with mesolimbic/mesocortical and incertohypothalamic dopaminergic neurons in male rats. *Hormones and Behavior, 62,* 505–514.

Sato, N., Tan, L., Tate, K., & Okada, M. (2015). Rats demonstrate helping behavior toward a soaked conspecific. *Animal Cognition, 18,* 1039–1047.

Singer, T., & Lamm, C. (2009). The social neuroscience of empathy. *Annals of the New York Academy of Sciences, 1156,* 81–96.

Singer, T., Seymour, B., O'Doherty, J., Kaube, H., Dolan, R. J., & Frith, C. D. (2004). Empathy for pain involves the affective but not sensory components of pain. *Science, 303,* 1157–1162.

Sivaselvachandran, S., Acland, E. L., Abdallah, S., & Martin, L. J. (2016). Behavioral and mechanistic insight into rodent empathy. *Neuroscience & Biobehavioral Reviews.* doi: 10.1016/j.neubiorev.2016.06.007 pii: S0149-7634(16)30063-X. [Epub ahead of print].

Taylor, S. E., Klein, L. C., Lewis, B. P., Gruenewald, T. L., Gurung, R. A., & Updegraff, J. A. (2000). Biobehavioral responses to stress in females: tend-and-befriend, not fight-or-flight. *Psychological Review, 107,* 411–429.

Zaki, J., Wager, T. D., Singer, T., Keysers, C., & Gazzola, V. (2016). The anatomy of suffering: understanding the relationship between nociceptive and empathic pain. *Trends in Cognitive Sciences, 20,* 249–259.

Zisook, S., Devaul, R. A., & Click, M. A. (1982). Measuring symptoms of grief and bereavement. *American Journal of Psychiatry, 139,* 1590–1593.

Relief From Stress Provided by Conspecifics: Social Buffering

Yasushi Kiyokawa

The University of Tokyo, Tokyo, Japan

INTRODUCTION

Humans commonly derive comfort and enjoyment from spending time with others. This is thought to be an underlying driver of friend-seeking behavior. One of the objective ways to evaluate such subjectively positive feelings may be to observe the reduction of stress responses to distressing stimuli in the presence of others (Kirschbaum, Klauer, Filipp, & Hellhammer, 1995; Thorsteinsson, James, & Gregg, 1998). Similarly, the presence of an affiliative conspecific(s) has been found to ameliorate stress responses in many nonhuman species, a phenomenon known as "social buffering" (Hennessy, Kaiser, & Sachser, 2009; Kiyokawa & Hennessy, 2018). The similarity of these phenomena suggests that humans and nonhuman animals may have similar positive feelings toward conspecific(s). This notion is supported by the observation that a rat released on a rat-free island swam to another island 400 m away, even though the rat-free island contained secure nesting places and abundant food (Russell, Towns, Anderson, & Clout, 2005). In this example, motivation to seek out and spend time with conspecifics appears to have been sufficiently strong to dive into the sea.

WHAT IS SOCIAL BUFFERING?

Although the presence of an affiliative conspecific has been reported to have a range of measurable effects, the term "social buffering" has been used ambiguously because the notion of social buffering is based on concepts such as "stress" and "social". In this chapter, I propose that the reduction of stress is a crucial component of social buffering. Hypothalamic-pituitary-adrenal

Neuronal Correlates of Empathy. http://dx.doi.org/10.1016/B978-0-12-805397-3.00011-5

(HPA) axis activation in response to distressing stimuli is currently the most widely accepted acute response reflecting the stress status of the animal. So are the changes in the activation pattern within neural circuits that drive HPA axis. The behavioral responses can serve as indirect indices of stress status, as they are governed by the neuronal defense system described above. However, it is impossible to use these acute responses as indices of stress when assessing forms of social buffering that develop more slowly. Although there seems to be no long-term alteration that is widely accepted as an index of stress, several physiological measures could serve as such an index. A detailed discussion regarding indices of stress in the context of social buffering was published previously (Kiyokawa & Hennessy, 2018).

One approach for clarifying appropriate measures is to divide social buffering into two experimental sequences depending on the presence or absence of a conspecific when a subject is exposed to distressing stimuli. In one sequence, the subject is exposed to distressing stimuli in the presence of a conspecific. In the other sequence, the subject is first exposed to distressing stimuli alone and then co-housed with a conspecific for days to weeks. These experimental sequences can be distinguished as "exposure-type" and "housing-type" social buffering, respectively. I have proposed the following definition for exposure-type social buffering: "phenomena in which stress of the subject is ameliorated when the subject is exposed to distressing stimuli along with a conspecific animal(s) (Kiyokawa, 2017)." Similarly, I defined housing-type social buffering as "phenomena in which recovery from adverse alterations induced by previously distressing stimuli is led by subsequent co-housing with a conspecific animal(s) (Kiyokawa, 2017)."

I assigned social buffering phenomena into two sequences simply because the measures used as indices of stress amelioration differ between exposure-type and housing-type. However, it is important to clarify, for the purpose of scientific discussions, the type of conspecific animal that induces social buffering, because data suggest that the neural mechanisms underlying social buffering differ depending on the type of conspecific animal. The terms "maternal buffering," "mate buffering," and "conspecific buffering," may be useful for specifying social buffering induced by a mother, mate, or same-sex or opposite-sex conspecifics without sexual relationships, respectively (Kiyokawa, 2017).

The remainder of this chapter focuses on conspecific buffering (i.e., social buffering induced by conspecifics without sexual relationships), beginning with a brief overview of current knowledge.

EXAMPLES OF EXPOSURE-TYPE SOCIAL BUFFERING

The first study reporting a possible social buffering phenomenon was conducted in 1955 (Davitz & Mason, 1955). A fear-conditioned or nonconditioned male subject rat was exposed to a visual conditioned stimulus

(CS) either individually, with a nonconditioned male rat or with a conditioned male rat. When activity was measured during an 11-min test period, the presence of a nonconditioned rat increased the activity of a conditioned subject, compared with a conditioned subject tested individually. Similarly, the presence of a conditioned rat tended to increase the activity of a conditioned subject. In contrast, the activity of a nonconditioned subject was similar when it was tested individually, with a nonconditioned rat or with a conditioned rat. Taken together, these findings suggest that the presence of a nonconditioned rat reduced the strength of the immobility exhibited by a conditioned subject. Although an index of stress was not provided in the experiment, it remains a pioneering study of social buffering.

After Davitz and Mason's (1955) groundbreaking study a wide variety of social buffering phenomena have been reported in a range of species. For example, rat pups with an immature HPA axis have been reported to emit ultrasonic "distress calls" when they are placed in a novel environment alone. In pups at the age of 12 days, the presence of a littermate drastically suppressed such distress calls (Hofer & Shair, 1978). Moreover, in animals with a functional HPA axis response, the presence of a conspecific(s) has been found to suppress HPA axis activation in response to a novel environment, as well as other behavioral and physiological responses. This was shown in chicks (Jones & Merry, 1988), mice (Klein et al., 2015), rats (File & Peet, 1980; Terranova, Cirulli, & Laviola, 1999), guinea pigs (Hennessy, Zate, & Maken, 2008), sheep (Lyons, Price, & Moberg, 1993), pigs (Kanitz, Hameister, Tuchscherer, Tuchscherer, & Puppe, 2014), common marmosets (Galvao-Coelho, Silva, & De Sousa, 2012), and rhesus monkeys (Winslow, Noble, Lyons, Sterk, & Insel, 2003). HPA axis activation in response to fearful stimuli such as aversive CS, predators, or humans, was also found to be ameliorated by the presence of a conspecific(s) in rats (Kiyokawa, Kikusui, Takeuchi, & Mori, 2004), goats (Lyons, Price, & Moberg, 1988), and squirrel monkeys (Stanton, Patterson, & Levine, 1985; Vogt, Coe, & Levine, 1981). In addition, the avoidance of predator odor was reduced by the presence of three conspecifics in rats (Bowen et al., 2013).

EXAMPLES OF HOUSING-TYPE SOCIAL BUFFERING

Housing-type social buffering has been less widely examined than exposure-type social buffering. A single social defeat was found to induce long-term alterations in a defeated rat, including weight loss, bradycardia and hypothermia during the dark phase, and increased anxiety when assessed with the elevated plus-maze test. However, these alterations were not observed when defeated rats were housed with a conspecific(s) after the defeat (de Jong, van der Vegt, Buwalda, & Koolhaas, 2005; Nakayasu & Ishii, 2008). In adult female Siberian hamsters, 2 h of a restraint procedure daily was found to delay wound healing, but this effect was blocked by housing with a female sibling (Detillion, Craft, Glasper, Prendergast, & DeVries, 2004). However, an experimental paradigm such as this may reflect the effect of isolation on HPA responsiveness to the restraint procedure, rather than the effect of social partners. Similarly, in California mice, housing with a conspecific was reported to enhance wound healing itself (Martin, Glasper, Nelson, & Devries, 2006). Due to the lack of an appropriate control group, it remains unclear whether co-housing enhanced wound healing, or if isolation delayed it. Further research is needed to clarify whether these phenomena constitute social buffering.

SOCIAL BUFFERING OF CONDITIONED FEAR RESPONSES IN RATS

In our experimental model, male Wistar subject rats were first fear-conditioned (or not) to an auditory CS. The next day, both conditioned and nonconditioned subjects were exposed to the auditory CS in a test box, either alone or with an accompanying rat. In our model, the accompanying rat was usually an unfamiliar male Wistar rat (Fig. 11.1). We found

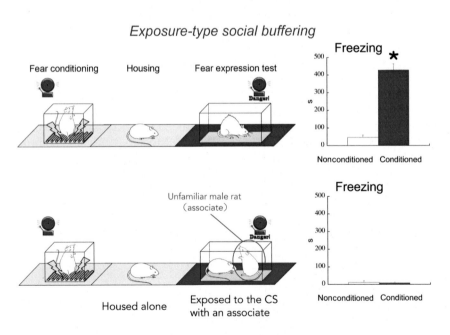

FIGURE 11.1 **Schematic diagram of the experiments assessing the effect of social buffering on conditioned fear responses.** In the lower panel: after being housed alone for 24 h after conditioning, fear-conditioned and nonconditioned subjects were exposed to the conditioned stimulus with an accompanying rat.

that the presence of an accompanying rat completely blocked freezing and HPA axis activation of a fear-conditioned subject in response to the auditory CS (Kiyokawa, Takeuchi, & Mori, 2007). Similar social buffering was observed between a female subject and an unfamiliar female rat (Ishii, Kiyokawa, Takeuchi, & Mori, 2016), suggesting that social buffering is a biologically important phenomenon in rats. Furthermore, we found that social buffering enhanced extinction of conditioned fear responses. When a fear-conditioned subject received social buffering during exposure to the CS, social buffering enhanced the efficacy of learning that the CS no longer predicted a foot shock. As a result, fear responses to the CS were extinguished when the fear-conditioned subject was again exposed to the CS on the following day, whereas the fear-conditioned subject that did not receive social buffering during the exposure still showed robust fear responses (Mikami, Kiyokawa, Takeuchi, & Mori, 2016).

Moreover, we found that an olfactory signal from another rat is sufficient to induce social buffering. When we tested rats in a box odorized by an unfamiliar rat, a fear-conditioned subject showed no fear response to the CS (Takahashi et al., 2013). In addition, this olfactory signal appeared to be volatile, because social buffering was induced even if a fear-conditioned subject was tested in an area that was separated from the odorized area by a punctured acrylic board partition that allowed the penetration of only volatile signals (Kiyokawa, Honda, Takeuchi, & Mori, 2014).

NEURAL MECHANISMS UNDERLYING SOCIAL BUFFERING OF CONDITIONED FEAR RESPONSES

We conducted several studies to analyze the neural mechanisms underlying social buffering. First, we sought to assess the role of the main olfactory system in detecting the volatile olfactory signal responsible for social buffering. We lesioned the main olfactory epithelium (MOE) of subjects, revealing that a lesion in the MOE blocked exposure-type social buffering (Kiyokawa, Takeuchi, Nishihara, & Mori, 2009). This finding suggests that the signal responsible for social buffering is detected in the MOE. Anatomical evidence that sensory neurons in the MOE send their projections only to the main olfactory bulb (MOB) suggests that all the signals are transmitted from the MOE to the MOB after detection. Given that the auditory CS evokes stress responses by activating the basolateral complex of the amygdala (BLA) (LeDoux, 2014), the signal responsible for social buffering appears to be transmitted from the MOB to the BLA, suppressing BLA activation. Because the MOB does not have direct projections to the BLA, certain brain structures must be intervening between the MOB and BLA. When we assessed the role of the regions receiving projections from the MOB, we found that a bilateral lesion of the posteromedial region of the olfactory peduncle (pmOP) specifically blocked social buffering (Kiyokawa, Wakabayashi, Takeuchi, & Mori, 2012). These results suggest that the pmOP is the primary site receiving the signal responsible for social buffering from the MOB. We further demonstrated that social buffering was blocked in a fear-conditioned subject with a unilateral pmOP lesion and contralateral, but not ipsilateral, BLA lesion (Kiyokawa et al., 2012). These findings suggest that the pmOP transmits the signal responsible for social buffering to ipsilateral BLA either directly or indirectly, and suppresses BLA activation. Further electrophysiological and immunohistochemical analyses have confirmed that lateral amygdala (LA) activation is suppressed during social buffering (Fuzzo et al., 2015; Kiyokawa et al., 2007; Kiyokawa & Honda et al., 2014; Takahashi et al., 2013), suggesting that the LA within the BLA is primarily responsible for receiving the signal. We also found that only the posterior complex of the anterior olfactory nucleus (AOP) within the pmOP showed increased c-Fos expression in fear-conditioned subjects when an accompanying rat induced social buffering (Takahashi et al., 2013). Taken together, these findings suggest that the olfactory signal responsible for social buffering is transmitted from the MOB to the pmOP, most likely the AOP region. The pmOP then suppresses LA activation during exposure-type social buffering. However, we simultaneously found no evidence that the presence of an accompanying rat increased c-Fos expression in the AOP of a non-conditioned subject (Takahashi et al., 2013). Therefore, the detection of an accompanying rat's olfactory signal does not automatically activate the AOP, implying the existence of an additional modifying system depending on the subjects' stress status.

SOCIAL BUFFERING OF CONDITIONED HYPERTHERMIA IN RATS

We have also established an experimental model for assessing housing-type social buffering of conditioned hyperthermia. Male rats were first fear-conditioned (or not) to an auditory CS. Conditioned and nonconditioned subjects were then housed either alone or with an unfamiliar male Wistar rat. One day after conditioning, subjects were exposed to the auditory CS in a test box alone (Fig. 11.2). When subjects were housed alone until CS exposure, fear-conditioned subjects showed stronger freezing, HPA axis activation, and hyperthermia compared with nonconditioned subjects. In contrast, hyperthermia responses and HPA axis activation were similar between conditioned and nonconditioned subjects when they were co-housed with another rat until CS exposure (Kiyokawa et al., 2007). These results cannot be explained by co-housing

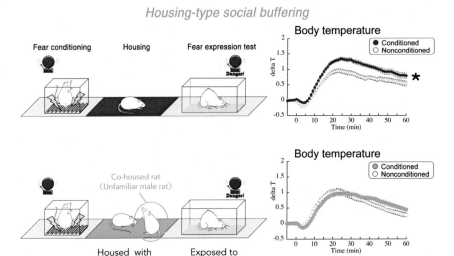

FIGURE 11.2   **Schematic diagram of the experiments assessing the effect of social buffering on conditioned hyperthermia.** In the lower panel: immediately after the conditioning procedure, fear-conditioned and nonconditioned subjects were co-housed with another rat for 24 h, then exposed to the conditioned stimulus alone.

serving as a chronic stressor and rendering a subject nonresponsive. We confirmed that co-housing did not induce HPA axis activation (Kiyokawa, Kodama, Takeuchi, & Mori, 2013). In addition, analyses of exposure-type social buffering revealed that encounters with another rat induced neither agonistic behaviors (Kiyokawa et al., 2004) nor HPA axis activation (Kiyokawa, Hiroshima, Takeuchi, & Mori, 2014). Therefore, co-housing with another rat does not appear to constitute a stressful event for rats. In addition, co-housing specifically ameliorated stress responses to the CS, rather than generally suppressing stress responses by reducing reactivity to distressing stimuli. Because the test box was a novel environment for the subject, even nonconditioned subjects showed hyperthermia responses. However, the intensity of hyperthermia responses was not affected by co-housing (Kiyokawa et al., 2007, 2013; Kiyokawa, Ishida, Takeuchi, & Mori, 2016; Kodama, Kiyokawa, Takeuchi, & Mori, 2011), suggesting that co-housing did not affect general reactivity to distressing stimuli. Taken together, these results suggest that co-housing induced social buffering of conditioned hyperthermia.

Subsequent analyses revealed several characteristics of social buffering. For example, we found that 12 h of co-housing was sufficient to induce social buffering (Kodama et al., 2011). In addition, co-housing commencing 6, 12, or 24 h after fear conditioning induced social buffering, suggesting that co-housing did not need to commence immediately after fear conditioning to induce social buffering (Kiyokawa et al., 2013, 2016; Kodama et al., 2011). Furthermore, social buffering was observed 12 h after the 12-h period of co-housing, which was extended to 48 h when the duration of co-housing was increased to 24-h (Kiyokawa et al., 2016). These results suggest that social buffering is sustained after a period of co-housing.

Although the signal that induces housing-type social buffering remains unclear, physical interactions between the subject and co-housed rat are not necessary to induce social buffering. We found that the placement of a wire mesh partition between a subject and co-housed rat did not block social buffering (Kiyokawa et al., 2013). Studies of exposure-type social buffering suggest that olfactory signals may also play an important role in the induction of housing-type social buffering.

NEURAL MECHANISMS UNDERLYING SOCIAL BUFFERING OF CONDITIONED HYPERTHERMIA

Along with behavioral, autonomic, and endocrine responses to the CS, fear-conditioned subjects have been found to exhibit increased c-Fos expression in the infralimbic cortex, central amygdala, LA, basal amygdala, and dorsomedial and dorsolateral periaqueductal grey when housed alone immediately after conditioning, until CS exposure. Housing-type social buffering was found to suppress increased c-Fos expression in the infralimbic cortex, central amygdala, basal amygdala, and dorsomedial periaqueductal grey in response to the CS (Kiyokawa et al., 2007). Therefore, co-housing appears to affect the brain, suppressing subsequent activation of these regions in response to the CS even if the subjects were exposed to the CS alone.

In one study, we found that co-housing commencing 24 h after conditioning induced social buffering (Kiyokawa et al., 2013), enabling analysis of the neural mechanisms inducing housing-type social buffering with minimal disturbance by the conditioning procedure *per se*. To examine the changes induced by co-housing, we observed c-Fos expression of fear-conditioned and nonconditioned subjects that were either housed alone or co-housed with another rat commencing 24 h after the conditioning. Examining c-Fos expression in response to co-housing revealed that c-Fos expression in the AOP was increased in fear-conditioned and co-housed subjects compared with fear-conditioned subjects and those housed alone. In contrast, nonconditioned and co-housed subjects showed similar level of c-Fos expression in the AOP, compared with nonconditioned subjects and those housed alone (Kiyokawa et al., 2013). These results suggest that the AOP plays an important role in the induction of housing-type social buffering. Given that the AOP activation in fear-conditioned subjects was also found during exposure-type social buffering, it is possible that exposure-type and housing-type social buffering are induced by similar neural mechanisms, in which the AOP plays an important role.

Although co-housing simultaneously increased c-Fos expression in the cingulate cortex, prelimbic cortex, infralimbic cortex, nucleus accumbens core and shell, the anterior part of the bed nucleus of the stria terminalis, the anterodorsal and posterodorsal parts of the medial amygdala, the posterolateral cortical amygdala, the LA, and the basal amygdala, these increments were observed both in fear-conditioned and co-housed subjects and nonconditioned and co-housed subjects (Kiyokawa et al., 2013). Further analyses are required to clarify the role of these regions in the induction of housing-type social buffering.

POSSIBLE PSYCHOLOGICAL FACTOR INDUCING SOCIAL BUFFERING IN RATS

Based on the findings of studies of exposure-type social buffering using several types of accompanying animals, I hypothesize that familiarity with an accompanying rat underlies social buffering effects in rats. Such familiarity primarily involves familiarity with the group to which the accompanying rat belongs. When the subject recognizes the accompanying rat as a member of the same group (e.g., a rat of a strain derived from the same colony), social buffering occurs. This familiarity effect enables an unfamiliar male Wistar accompanying rat to induce social buffering in a male Wistar subject rat (Kiyokawa et al., 2007). At the same time, interaction with an unfamiliar male guinea pig did not induce social buffering because the Wistar subject rat recognized it as an out-group member of a different species (Kiyokawa et al., 2009). This hypothesis is also consistent with our recent finding that an unfamiliar male Wistar, Sprague–Dawley, Long–Evans, and Lewis rat induced a similar level of social buffering in a Wistar male subject rat (Nakamura, Ishii, Kiyokawa, Takeuchi, & Mori, 2016). Given that these strains of rats are descended from Wistar rats (Lindsey & Baker, 2006), the subject rat presumably evaluated them as members of the same group, and sensed familiarity. In contrast, neither an unfamiliar Fischer 344 (F344) nor a Brown Norway rat induced social buffering in a Wistar subject rat (Nakamura et al., 2016). Because these rats have been bred independently of Wistar rats (Lindsey & Baker, 2006), the subject did not sense familiarity with these accompanying rats.

Prior interaction with a given individual appears to strengthen the familiarity effect. For example, a cagemate Wistar rat that had been co-housed with a Wistar subject rat for 3 weeks induced stronger social buffering effects than an unfamiliar Wistar rat (Kiyokawa & Honda et al., 2014). Because the observed responses represent the residual stress responses induced by the distressing stimuli after suppression by social buffering, the subject rat might have stronger familiarity with the cagemate than with an unfamiliar rat. Therefore, a subject rat appears to have additional familiarity with an accompanying rat when the subject recognizes the accompanying rat as its own cagemate.

SUMMARY AND IMPLICATIONS OF SOCIAL BUFFERING PHENOMENA

This chapter summarized our knowledge regarding social buffering induced by conspecifics with nonsexual relationships. I first proposed to divide the phenomena into two types: (1) In exposure-type social buffering, acute stress responses of the subject are mitigated when the subject is exposed to distressing stimuli in the presence of an affiliative conspecific. (2) In housing-type social buffering, recovery from adverse alterations induced by the previous distressing stimuli is led by subsequent co-housing with an affiliative conspecific. After this classification, I summarized current knowledge about each type of social buffering. Finally, I briefly introduced our recent findings regarding exposure-type social buffering of conditioned fear responses, and housing-type social buffering of conditioned hyperthermia in rats. These findings suggest that social buffering is induced when the subject is familiar with accompanying animals.

Because our knowledge on these phenomena is still developing, many interesting questions remain to be answered. For example, we found that an olfactory signal from an unfamiliar Wistar rat induces social buffering in a Wistar subject rat (Kiyokawa & Honda et al., 2014; Takahashi et al., 2013). This implies that a subset of molecules represent the Wistar strain and induce social buffering in the subject. In addition, AOP activation has been suggested to suppress the LA during social buffering (Fuzzo et al., 2015; Kiyokawa et al., 2012; Takahashi et al., 2013). However, it remains unclear whether the AOP directly suppresses the LA or indirectly suppresses the LA by activating another brain structure that can inhibit LA activity. Furthermore, future studies are required to test the hypothesis that familiarity with an accompanying rat induces social buffering. In particular, it is unclear whether in-group familiarity and familiarity with the cagemate operate on the same axis. Finally, sensing in-group familiarity would presumably involve a subject rat comparing its own features with that of an accompanying rat. However, at this time, it remains unclear how a subject establishes such recognition. We are currently investigating these questions.

Social buffering appears to be an important biological component of gregariousness in animals. The ability to receive social buffering may have been acquired during evolution. The reduction of stress increases survival rate of animals in proximity to conspecifics, increasing their fitness. If a species has evolved sufficiently complex emotional capabilities, reduction of stress may simultaneously evoke a sense of relief and/or affiliation, motivating them to increase the time spent with conspecifics. As a result, the number of animals living in a group would increase. Living in groups brings additional benefits, such as protection from physical environmental factors and predators, which can enhance fitness. The genetic mechanisms responsible for social buffering would consequently prevail among conspecifics, resulting in the development of gregariousness in the species. Therefore, social buffering provides a valuable experimental model for understanding the neurobiology of positive emotions such as the sense of relief and/or affiliation, as well as the biological causes of gregariousness.

Acknowledgments

I would like to thank Dr. Michael Hennessy and Akiko Ishii for their helpful suggestions on the manuscript. I would also like to thank the late Professor Yuji Mori for giving me the opportunity to conduct social buffering research as a graduate student, postdoc, and assistant professor.

References

Bowen, M. T., Kevin, R. C., May, M., Staples, L. G., Hunt, G. E., & McGregor, I. S. (2013). Defensive aggregation (huddling) in Rattus norvegicus toward predator odor: Individual differences, social buffering effects and neural correlates. *PLoS One, 8*, e68483.

Davitz, J. R., & Mason, D. J. (1955). Social facilitated reduction of a fear response in rats. *Journal of Comparative and Physiological Psychology, 48*, 149–156.

de Jong, J. G., van der Vegt, B. J., Buwalda, B., & Koolhaas, J. M. (2005). Social environment determines the long-term effects of social defeat. *Physiology & Behavior, 84*, 87–95.

Detillion, C. E., Craft, T. K., Glasper, E. R., Prendergast, B. J., & DeVries, A. C. (2004). Social facilitation of wound healing. *Psychoneuroendocrinology, 29*, 1004–1011.

File, S. E., & Peet, L. A. (1980). The sensitivity of the rat corticosterone response to environmental manipulations and to chronic chlordiazepoxide treatment. *Physiology & Behavior, 25*, 753–758.

Fuzzo, F., Matsumoto, J., Kiyokawa, Y., Takeuchi, Y., Ono, T., & Nishijo, H. (2015). Social buffering suppresses fear-associated activation of the lateral amygdala in male rats: Behavioral and neurophysiological evidence. *Frontiers in Neuroscience, 9*, 99.

Galvao-Coelho, N. L., Silva, H. P., & De Sousa, M. B. (2012). The influence of sex and relatedness on stress response in common marmosets (Callithrix jacchus). *American Journal of Primatology, 74*, 819–827.

Hennessy, M. B., Zate, R., & Maken, D. S. (2008). Social buffering of the cortisol response of adult female guinea pigs. *Physiology & Behavior, 93*, 883–888.

Hennessy, M. B., Kaiser, S., & Sachser, N. (2009). Social buffering of the stress response: Diversity, mechanisms, and functions. *Frontiers in Neuroendocrinology, 30*, 470–482.

Hofer, M. A., & Shair, H. (1978). Ultrasonic vocalization during social interaction and isolation in 2-weeek-old rats. *Developmental Psychobiology, 11*, 495–504.

Ishii, A., Kiyokawa, Y., Takeuchi, Y., & Mori, Y. (2016). Social buffering ameliorates conditioned fear responses in female rats. *Hormones and Behavior, 81*, 53–58.

Jones, R. B., & Merry, B. J. (1988). Individual or paired exposure of domestic chicks to an open field: Some behavioural and adrenocortical consequences. *Behavioural Processes, 16*, 75–86.

Kanitz, E., Hameister, T., Tuchscherer, M., Tuchscherer, A., & Puppe, B. (2014). Social support attenuates the adverse consequences of social deprivation stress in domestic piglets. *Hormones and Behavior, 65*, 203–210.

Kirschbaum, C., Klauer, T., Filipp, S. H., & Hellhammer, D. H. (1995). Sex-specific effects of social support on cortisol and subjective responses to acute psychological stress. *Psychosomatic Medicine, 57*, 23–31.

Kiyokawa, Y. (2017). Social odors: Alarm pheromones and social buffering. *Current Topics in Behavioral Neurosciences, 30*, 47–65.

Kiyokawa, Y., & Hennessy, M. B. (2018). Comparative studies of social buffering: A consideration of approaches, terminology, and pitfalls. *Neuroscience & Biobehavioral Reviews*.

Kiyokawa, Y., Kikusui, T., Takeuchi, Y., & Mori, Y. (2004). Partner's stress status influences social buffering effects in rats. *Behavioral Neuroscience, 118*, 798–804.

Kiyokawa, Y., Takeuchi, Y., & Mori, Y. (2007). Two types of social buffering differentially mitigate conditioned fear responses. *European Journal of Neuroscience, 26*, 3606–3613.

Kiyokawa, Y., Takeuchi, Y., Nishihara, M., & Mori, Y. (2009). Main olfactory system mediates social buffering of conditioned fear responses in male rats. *European Journal of Neuroscience, 29*, 777–785.

Kiyokawa, Y., Wakabayashi, Y., Takeuchi, Y., & Mori, Y. (2012). The neural pathway underlying social buffering of conditioned fear responses in male rats. *European Journal of Neuroscience, 36*, 3429–3437.

Kiyokawa, Y., Kodama, Y., Takeuchi, Y., & Mori, Y. (2013). Physical interaction is not necessary for the induction of housing-type social buffering of conditioned hyperthermia in male rats. *Behavioural Brain Research, 256*, 414–419.

Kiyokawa, Y., Hiroshima, S., Takeuchi, Y., & Mori, Y. (2014). Social buffering reduces male rats' behavioral and corticosterone responses to a conditioned stimulus. *Hormones and Behavior, 65*, 114–118.

Kiyokawa, Y., Honda, A., Takeuchi, Y., & Mori, Y. (2014). A familiar conspecific is more effective than an unfamiliar conspecific for social buffering of conditioned fear responses in male rats. *Behavioural Brain Research, 267*, 189–193.

Kiyokawa, Y., Ishida, A., Takeuchi, Y., & Mori, Y. (2016). Sustained housing-type social buffering following social housing in male rats. *Physiology & Behavior, 158*, 85–89.

Klein, B., Bautze, V., Maier, A. M., Deussing, J., Breer, H., & Strotmann, J. (2015). Activation of the mouse odorant receptor 37 subsystem coincides with a reduction of novel environment-induced activity within the paraventricular nucleus of the hypothalamus. *European Journal of Neuroscience, 41*, 793–801.

Kodama, Y., Kiyokawa, Y., Takeuchi, Y., & Mori, Y. (2011). Twelve hours is sufficient for social buffering of conditioned hyperthermia. *Physiology & Behavior, 102*, 188–192.

LeDoux, J. E. (2014). Coming to terms with fear. *Proceedings of the National Academy of Sciences of the United States of America, 111*, 2871–2878.

Lindsey, J. R., & Baker, H. J. (2006). Historical foundations. In M. A. Suckow, S. H. Weisbroth, & C. L. Franklin (Eds.), *The laboratory rat* (2nd ed.). New York: Elsevier Academic Press pp. 1–52.

Lyons, D. M., Price, E. O., & Moberg, G. P. (1988). Social modulation of pituitary-adrenal responsiveness and individual differences in behavior of young domestic goats. *Physiology & Behavior, 43*, 451–458.

Lyons, D. M., Price, E. O., & Moberg, G. P. (1993). Social grouping tendencies and separation-induced distress in juvenile sheep and goats. *Developmental Psychobiology, 26*, 251–259.

Martin, L. B., 2nd, Glasper, E. R., Nelson, R. J., & Devries, A. C. (2006). Prolonged separation delays wound healing in monogamous California mice, Peromyscus californicus, but not in polygynous white-footed mice *P. leucopus*. *Physiology & Behavior, 87*, 837–841.

Mikami, K., Kiyokawa, Y., Takeuchi, Y., & Mori, Y. (2016). Social buffering enhances extinction of conditioned fear responses in male rats. *Physiology & Behavior, 163,* 123–128.

Nakamura, K., Ishii, A., Kiyokawa, Y., Takeuchi, Y., & Mori, Y. (2016). The strain of an accompanying conspecific affects the efficacy of social buffering in male rats. *Hormones and Behavior, 82,* 72–77.

Nakayasu, T., & Ishii, K. (2008). Effects of pair-housing after social defeat experience on elevated plus-maze behavior in rats. *Behavioural Processes, 78,* 477–480.

Russell, J. C., Towns, D. R., Anderson, S. H., & Clout, M. N. (2005). Intercepting the first rat ashore. *Nature, 437,* 1107.

Stanton, M. E., Patterson, J. M., & Levine, S. (1985). Social influences on conditioned cortisol secretion in the squirrel monkey. *Psychoneuroendocrinology, 10,* 125–134.

Takahashi, Y., Kiyokawa, Y., Kodama, Y., Arata, S., Takeuchi, Y., & Mori, Y. (2013). Olfactory signals mediate social buffering of conditioned fear responses in male rats. *Behavioural Brain Research, 240,* 46–51.

Terranova, M. L., Cirulli, F., & Laviola, G. (1999). Behavioral and hormonal effects of partner familiarity in periadolescent rat pairs upon novelty exposure. *Psychoneuroendocrinology, 24,* 639–656.

Thorsteinsson, E. B., James, J. E., & Gregg, M. E. (1998). Effects of video-relayed social support on hemodynamic reactivity and salivary cortisol during laboratory-based behavioral challenge. *Health Psychology, 17,* 436–444.

Vogt, J. L., Coe, C. L., & Levine, S. (1981). Behavioral and adrenocorticoid responsiveness of squirrel monkeys to a live snake: Is flight necessarily stressful? *Behavioral and Neural Biology, 32,* 391–405.

Winslow, J. T., Noble, P. L., Lyons, C. K., Sterk, S. M., & Insel, T. R. (2003). Rearing effects on cerebrospinal fluid oxytocin concentration and social buffering in rhesus monkeys. *Neuropsychopharmacology, 28,* 910–918.

Helping Behavior in Rats

Inbal Ben-Ami Bartal, Peggy Mason

University of Chicago, Chicago, IL, United States

Empathy refers to the communication of affect or experience between two individuals. Nonhuman animals who lack articulated speech were long thought immune from such communication. However, de Waal's pioneering work in primates shattered that myth (de Waal, 2012). More recently, work has demonstrated empathy between individual rodents. In this chapter, we focus on one possible outward manifestation of empathy, helping, and on rodents.

HELPING IS AN OBSERVABLE OUTCOME OF EMPATHY

In 1959 Russell Church, a psychologist at Brown, reported that rats gave up the chance to receive food rather than inflict pain on another rat (Church, 1959). In this experiment, rats were trained to press a lever to receive a food reward. After learning this task, a twist was added. If rats pressed the lever, they would still get the reward, but a conspecific, visible from across a barrier, would also receive a painful electric shock. Rats stopped pressing the lever. Moreover, rats who had previously been shocked themselves stopped for a longer time than did naive rats. The results appeared clear-cut, and Church interpreted them as evidence that rats acted out of empathy for another rat. This work was met with controversy and criticism, mostly because an easy alternative explanation was that rats simply froze in fear as they witnessed another rat's distress. The rats in Church's study did not *actively* do anything to benefit another. Moreover, most rats were immobile for only a few seconds, after which they went right back to pressing the lever, shocks and all. Church's work did not make a large immediate impact, not only because of methodological concerns, but also because the field, still immersed in Skinnerian behaviorism, was not receptive to the idea of feelings in rats.

Neuronal Correlates of Empathy. http://dx.doi.org/10.1016/B978-0-12-805397-3.00012-7

Church's experiment was not perfect (no experiment ever is) but none-theless tells us something important: rats are sensitive to the emotions of other rats. Even the interpretation that rats simply freeze in fear at the sight of a distressed conspecific necessarily means that rats are affected by the experiences of other rats. Essentially, Church's report showed that distress is communicated from a shocked rat to an observer rat. Such a transfer of affect between individuals, often termed emotional contagion, has been well documented in rodents, as is elaborated upon elsewhere in this book. Here we focus on one potential behavioral sequel to emotional contagion, namely helping. By definition, helping, which is acting for the benefit of other individuals, is a prosocial act. Individuals that perform this action are acting prosocially; schadenfreude occurs with a decrease in the frequency of helping and represents helping's antisocial opposite.

RATS HELP ANOTHER IN DISTRESS

Until 2011 the evidence for rodents actively helping each other was mostly anecdotal. Sparse empirical evidence included an experiment inspired by Church's work, in which rats pressed a lever to bring down a distressed rat who was suspended midair (Rice & Gainer, 1962). If the suspended rat did not squirm enough, the investigators poked it. Although Rice and Gainer's results were convincing, the paradigm never caught on, even before today's sensibility to the ethical treatment of laboratory animals. In another paradigm, rats swam underwater to fetch food for rats on the other side (Grasmuck & Desor, 2002; Krafft, Colin, & Peignot, 2010). Finally, quid pro quo and generalized reciprocal food sharing both occur in rats. Rats push food over to another rat who has provided them with food in the past (direct reciprocity) or to stranger rats if they have received food repeatedly from other strangers (generalized reciprocity; Dolivo, Rutte, & Taborsky, 2016; Rutte & Taborsky, 2007). These latter experiments certainly involve behavior that can be construed as cooperative, prosocial, and even "nice." However, it is less clear that food-sharing in the absence of need (e.g., food deprivation) or distress qualifies as help. In any case, as with Church's study, these studies did not receive the attention they merited, possibly not only due to an enduring bias toward behaviorism but also potentially due to the complexity of the experimental setups.

Looking at the available literature in 2007, we believed that there was a need for an experimentally tractable model of rodent helping. We wanted the model to be simple, easy to grasp, and to provide a reliable answer to the question of whether a rat was motivated to actively help another rat in distress. As multiple studies, including those discussed previously, had demonstrated emotional contagion in mice and rats, we focused our attention on testing whether rats could use another's distress as motivation

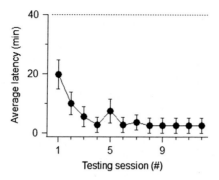

FIGURE 12.1 Rats (n = 16) were placed in a restrainer with a reversed door that could only be opened from the inside. The average latency (± SEM) until trapped rats released themselves decreased rapidly across the days of testing.

for active helping. Our goal was simple: give rats a "rat-sized" challenge that would enable them to *do something* to help another rat *if* they were so motivated.

Trapping rats inside a plastic tube or "restrainer" has been used for decades as a stressor to study the physiology of stress. Although no tissue injury occurs and rats do not express pain behaviors, trapped rats demonstrate signs of distress. They mount a stress response, including elevations in corticosterone and sympathetic activation. We have found that, given the opportunity, rats escape at short latency from a restrainer (Fig. 12.1). Thus, placing a rat inside a restrainer is an ethical method for generating a state of moderate distress.

Our interest was not in a rat who directly experiences restraint stress but rather in a rat who could only feel restraint stress "vicariously." We wanted to know whether a rat with no direct source of distress could be motivated through social communication to help a different rat who is trapped and therefore distressed. To ask this question we fashioned a custom restrainer door that could only be opened from the outside and thus only by the free rat. To let out a trapped cagemate, a free rat simply has to nudge up the restrainer door with his snout. This simple-sounding idea was remarkably difficult for Sprague Dawley rats to execute. We humans understand Plexiglas, doors, and counterweights. Rats do not have such an understanding before physically engaging with the objects. What we instantly recognize to be a door is not immediately identified as such by rats. Our challenge then was to enable the rats to discover their own capacity to open the restrainer door and release the trapped rat.

Typically, rats are taught operant behaviors such as pressing a lever to receive a food reward by being subjected to tens or hundreds of trials a day over the course of days to weeks. They are trained to press a lever for food before being tested on the target behavior of interest. However, we did not want food involved in any way because we worried that food-trained rats might ultimately be motivated to open the restrainer door in hope of receiving a *remembered* food reward. We were interested in rats that were motivated to release a trapped rat solely by a previous experience of having opened the door. We did not want even the remotest possibility that rats would associate door-opening with any other reward.

Inspired by the Morris water maze, which tracks spatial learning over the course of several days without providing explicit training *per se*, we placed our rats in an arena with a trapped cagemate repeatedly for 12 days, for 1 h each day. In essence, we wanted rats *to teach themselves* the task of opening the door just as rats do to learn the location of a platform in a sea of opaque water. To extend this analogy one more step, we reasoned that a rat's first door opening is as much of a lucky accident as is a rat's first encounter with a hidden platform. A single opening is thus not of particular interest. Instead, we wanted to understand the rat's experience of that single opening. We reasoned that if the experience of opening the door and releasing the trapped rat is a positive one, "rewarding" in today's parlance, then the rat would want to do it again. This type of behavioral reinforcement is exactly what we observed (Fig. 12.2). Across multiple sessions, most rats teach themselves to open the door and then do so consistently, thereby releasing their cagemate from the restrainer (Ben-Ami Bartal, Decety, & Mason, 2011).

Rats open the restrainer door at increasing frequencies and progressively shorter latencies on successive days of testing (Ben-Ami Bartal et al., 2011). Of great importance is the fact that this putative "helping"

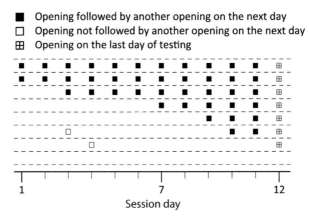

FIGURE 12.2 Door-opening behavior is illustrated for a cohort of eight pairs of Sprague-Dawley rats. Each row represents the behavior of a different free rat. Most occurrences of door-openings are followed by door-opening on the subsequent day (*filled symbols*). *Source: Data taken from Ben-Ami Bartal, I., Rodgers, D.A., Bernardez Sarria, M.S., Decety J., & Mason P. (2014). Pro-social behavior in rats is modulated by social experience. eLife, 3, e01385.*

behavior requires action not inactivity. In addition, door-opening behavior for a trapped rat has ecological validity, closely resembling rescuing behavior that may occur after burrow cave-ins that occur in the wild. It is also important that helping procures no immediate external benefit to the helper.

A RAT'S MOTIVATION TO RELEASE A TRAPPED CONSPECIFIC

Identifying the free rat's motivation to release a trapped rat is key to interpreting this behavior. When tested with cagemates, albino rats require 3–6 days to learn to open the restrainer. In contrast, rats placed in an arena with either an empty restrainer or a restrainer containing a toy rat only open the restrainer sporadically and never do so consistently. Consistent and short latency door openings, which are signs of reinforcement, are not observed in control conditions (Ben-Ami Bartal et al., 2011). Thus, results from these control conditions rule out the possibility that rats open the restrainer because they are curious about the restrainer itself or the presence of an object inside. Only rats tested with a trapped cagemate circle the restrainer, bite it, dig under it, and persist in these efforts throughout the length of testing sessions and over multiple days. Once a rat opens the restrainer door, he opens the door on the following day with a probability greater than 90%. In sum, the *presence of the trapped rat* is the impetus for free rats to open the restrainer door.

Before accepting that door opening is motivated by a rodent version of empathy, we had to address the legitimate possibility that the free rat opens the door simply because he wants to play with the trapped rat. To test this entirely reasonable possibility, we built a modified setup with two attached arenas (Ben-Ami Bartal et al., 2011). The restrainer was inserted into a hole in the wall between the arenas and the door was placed such that, when opened, the trapped rat was released into an adjacent arena. Thus trapped rats were released into a separate arena from the arena containing the free rat who had opened the door. This configuration prevented immediate social contact between the free rat and his newly released cagemate. Over several months of testing, rats persisted in opening the restrainer despite the lack of postdoor-opening contact with the released cagemate. Yet door-opening was extinguished when an empty restrainer was placed in the setup. In other words, free rats opened the door only if there was a trapped cagemate present in the restrainer and even without being able to gain access to the released rat. The finding that the motivation for releasing a trapped rat is independent of immediate social contact has recently been confirmed (Ben-Ami Bartal et al., 2016; Sato, Tan, Tate, & Okada, 2015).

Additional experiments revealed more about the circumstances when rats were willing to help. For instance we found that the value of helping for rats is on par with the value of chocolate chips (Ben-Ami Bartal et al., 2011). We also observed that helping does not occur at any cost. Rats are willing to tread through a pool of water, but not of oil, to release a trapped rat (unpublished results). Rats are also selective in *who* they will help (Ben-Ami Bartal, Rodgers, BernardezSarria, Decety, & Mason, 2014). Their social selectivity is based solely on familiarity with rat type and not on individual familiarity: rats open for strangers, as well as for known individuals. Remarkably, rats help any rat of a familiar type but do not help rats of an unfamiliar type, even when that unfamiliar type matches their own biological type.

AN AFFECTIVE MOTIVATION TO RELEASE A TRAPPED CONSPECIFIC

When free rats open the door to a restrainer, a trapped rat is released. Thus, the released rat benefits from the free rat's action. Moreover, being released relieves the distress experienced when trapped. Thus door opening on the part of free rats terminates the distress of and directly benefits the erstwhile-trapped rat. An other-oriented behavior that relieves the distress of another has the hallmarks of intentional helping. Yet, ants release other ants from snares, actions that benefit the released ants and also appear to fulfill the criteria for "helping" behavior. Does the fact that ants act in ways that benefit other ants diminish from the helping actions of rats? In this section, we argue that the mechanisms supporting conspecific-benefiting actions in rats and humans are similar, whereas the mechanisms supporting such behavior in ants are (1) unknown, and (2) unlikely to resemble mechanisms found in mammals.

The affective machinery at the root of the rodent response to another's distress is grossly analogous to that of humans. The mammalian limbic system, most particularly the amygdala, is responsible for tying emotional significance to particular stimuli. By virtue of experience, stimuli are connected to specific emotions through a learning process. This type of learning aids animals by favoring approach toward rewarding stimuli and avoidance of adverse circumstances. Rats, humans, and other mammals employ the same set of neurotransmitters, including oxytocin, vasopressin, dopamine, and serotonin, in fundamentally similar ways. Mechanisms that lead to activation of the hypothalamic adrenocortical axis and the consequences thereof are also shared across the mammalian clade. These similarities are the main reason that rodent models are ubiquitous in biomedical research and particularly in behavioral research.

The demonstration of emotional contagion in rodents shows that rodents have the biological capacity to feel the distress of others (Panksepp & Lahvis, 2011). This raises the obvious possibility that rats help because they are distressed by the distress of another and that they act to terminate the other's distress. We wanted to directly test if emotional contagion of another's emotion is the driving force for helping another in distress. Therefore we administered a drug that prevented free rats from feeling distressed, and then examined their response to a trapped rat (Ben-Ami Bartal et al., 2016). The drug we used is a benzodiazepine called midazolam, which is commonly prescribed to humans to reduce anxiety. Remarkably, midazolam-treated rats did not open the restrainer door for trapped rats. To determine if rats were not opening the restrainer because of the sedating side effects of the drug, another group of rats was tested with a restrainer containing chocolate chips. Midazolam-treated rats readily opened the restrainer door and consumed the chips inside. These experiments demonstrate that a transfer of distress between the trapped and free rats is required for helping to occur. Furthermore, they show that the free rat's motivation to release a trapped cagemate is based in emotion. Blocking the free rat's access to the full range of affect blocks helping.

PROSOCIAL ACTIONS ABOUND IN RODENTS

Since we published our findings in 2011, several other demonstrations of prosocial behavior in rodents have emerged. Of greatest relevance, Sato and colleagues examined the behavior of subject rats faced with a target rat trapped in a pool of water rather than in a Plexiglas restrainer (Sato et al., 2015). Subject rats placed in a (dry) compartment adjacent to the pool opened a door that allowed the wet target rat to escape from the water. An advantage to this paradigm is that the pool compartment is only stressful if filled with water. Correspondingly, subject rats only opened the door when the pool was filled, directly demonstrating that subject rats were not opening the door simply to play with target rats.

A popular paradigm in monkeys is food sharing. Monkeys are given the choice of receiving a food reward only for themselves, or in concert with another monkey also receiving the same food. In both situations, the subject monkey receives the same food. Thus, monkeys are asked to choose between a "selfish" option, where only they receive food, and a "prosocial" option, where another individual is also rewarded at no cost to them. This paradigm was adapted for rats independently by two groups, who happily observed nearly identical results. In essence, rats prefer the prosocial choice to the selfish choice (Hernandez-Lallement, van Wingerden, Marx, Srejic, & Kalenscher, 2015; Márquez, Rennie, Costa, & Moita, 2015). This preference critically depends on demonstration of food-seeking behavior on the part of the target rat (Márquez et al., 2015). This was another important demonstration that rats are sensitive to the well being of other, nonrelated individuals.

In sum, mounting evidence shows that rats are emotionally motivated to intentionally benefit others. Actions may take the form of *help* as occurs when trapped rats are freed from their distress-producing circumstances or they may be generous acts that provide food to another without either benefit or cost to the self.

EMPATHY IS A MORALLY NEUTRAL TERM

Our and others' reports on rat empathy and helping are welcome news to the general public, as they are interpreted as news from biology to feel good about. Yet, it is important to remember the disadvantages of empathy and helping. Helping is resource depleting, both physically and mentally. A person who acted on every feeling of empathic concern while walking through a modern urban area would run out of money and never ever reach home, eventually collapsing in exhaustion. Attending and caring for all individuals in need is not sustainable on either an individual or evolutionary level.

It is also important to remember that empathy may drive behavior that appears caring but is in fact unwanted by the recipient. A person may "help" to build social capital with the aim of creating indebtedness in the target individual toward himself or herself. Thus, empathy may not always be a positive motivator. This is well illustrated by a consideration of the trolley car paradigm. Imagine that a railway car is heading straight for five people tied to the railroad track. You can pull a switch so that the train is diverted to a track upon which only one person is tied. A common "solution" to this dilemma is to pull the switch and divert the train to reduce the death toll from five to one. Now consider that a single loved one, a child or parent, for example, is tied to the diversion track while five strangers are tied down on the train's present track. In this scenario, empathy for the loved one would favor not pulling the switch and thus letting five strangers die rather than allowing one loved individual to perish. Thus, empathy can motivate acts that run counter to moral endpoints.

OVEREMPHASIS OF THE POSITIVE SIDE OF EMPATHY RETARDS SCIENTIFIC PROGRESS

As previously explained, empathy can power behavior that either benefits others (prosocial) or harms others (antisocial). Empathy is a neutral phenomenon that simply refers to the communication of affect or experience between two individuals. Popular emphasis is on empathy as a motivator for helping and other prosocial behaviors. However, empathy can also motivate highly undesirable phenomena. For example, in mob violence or rioting, the frequency of destructive and injurious behavior is greater for a group than would be expected from the same number of individuals acting independently. As another example, empathy can be used to personalize and exacerbate another's distress as occurs in torture. Thus empathy can drive either prosocial helping or highly antisocial activities.

Empathy is widely lauded as desirable by politicians, activists, celebrities, and the general public. When he nominated Sonya Sotomayor for the Supreme Court of the United States, President Barack Obama explained that he saw the "quality of empathy … as an essential ingredient for arriving at just decisions and outcomes." Websites with names such as, cultureofempathy.com, rootsofempathy.org, and compassionsociety.org promote empathy as a way to combat a variety of societal ills and break down distrust between people. The popular idea that empathy can cure the social ailments of modern society has seduced researchers, leading to a biased approach that may be retarding scientific progress.

Empathy, along with cooperation and pair-bonding, exemplify the typical topics of social neuroscience. These topics are not neutral but rather are nearly universally viewed as desirable. Matusall and colleagues (2011) write that, "today's characterization … of the term "social" often sets priorities on "positive" issues like cooperation, empathy, care," and so on, leading to a "contamination" of biology with societal ideals. As an example of contamination between "social" and "prosocial," consider the biologically relevant (i.e., not human-specific) definitions of *social* in the Merriam-Webster Dictionary: (1) "tending to form cooperative and interdependent relationships with others;" (2) "living and breeding in more or less organized communities." The latter applies particularly to insects and the former to the broader animal kingdom. These definitions belie the common bias that sociality means prosociality. However, territoriality, aggression, and bullying are all implicitly social in that they are actions that require at least two actors. They cannot occur in a world of one individual. There is no justification for carving out the positive end of the interactive spectrum as social and excluding interactive behaviors that are either neutral or antagonistic. Scientific examination of empathic interactions and socially mediated behaviors in an unbiased manner represents a major opportunity for the future.

Acknowledgments

Discussions with Haozhe Shan and his comments on the manuscript are gratefully acknowledged.

References

Ben-Ami Bartal, I., Decety, J., & Mason, P. (2011). Empathy and pro-social behavior in rats. *Science, 334,* 1427–1431.

Ben-Ami Bartal, I., Rodgers, D. A., Bernardez Sarria, M. S., Decety, J., & Mason, P. (2014). Prosocial behavior in rats is modulated by social experience. *eLife, 3,* e01385.

Ben-Ami Bartal, I., Shan, H., Molasky, N. M., Murray, T. M., Williams, J. Z., Decety, J., et al. (2016). Anxiolytic treatment impairs helping behavior in rats. *Frontiers in Psychology, 8*(7), 850.

Church, R. (1959). Emotional reactions of rats to the pain of others. *Journal of Comparative Physiological Psychology, 52,* 132–134.

de Waal, F. B. M. (2012). Empathy in primates and other mammals. In J. Decety (Ed.), *Empathy: From bench to bedside*. Cambridge, MA: The MIT Press pp. 87–106.

Dolivo, V., Rutte, C., & Taborsky, M. (2016). Ultimate and proximate mechanisms of reciprocal altruism in rats. *Learning & Behavior, 44,* 223–226.

Grasmuck, V., & Desor, D. (2002). Behavioural differentiation of rats confronted to a complex diving-for-food situation. *Behavioural Processes, 58,* 67–77.

Hernandez-Lallement, J., van Wingerden, M., Marx, C., Srejic, M., & Kalenscher, T. (2015). Rats prefer mutual rewards in a prosocial choice task. *Frontiers in Neuroscience, 8,* 1–9.

Krafft, B., Colin, C., & Peignot, P. (2010). Diving-for-food: A new model to assess social roles in a group of laboratory rats. *Ethology, 96,* 11–23.

Márquez, C., Rennie, S. M., Costa, D. F., & Moita, M. A. (2015). Prosocial choice in rats depends on food-seeking behavior displayed by recipients. *Current Biology, 25,* 1736–1745.

Matusall, S., Kaufmann, I. M., & Christen, M. (2011). The emergence of social neuroscience as an academic discipline. In J. Decety, & J. T. Cacciopo (Eds.), *The Oxford handbook of social neuroscience* (pp. 9–27). New York: Oxford University Press.

Panksepp, J. B., & Lahvis, G. P. (2011). Rodent empathy and affective neuroscience. *Neuroscience and Biobehavioral Reviews, 35,* 1864–1875.

Rice, G. E., & Gainer, P. (1962). Altruism in the albino rat. *Journal of Comparative and Physiological Psychology, 55,* 123–125.

Rutte, C., & Taborsky, M. (2007). Generalized reciprocity in rats. *PLoS Biology, 5,* e196.

Sato, N., Tan, L., Tate, K., & Okada, M. (2015). Rats demonstrate helping behavior toward a soaked conspecific. *Animal Cognition, 18,* 1039–1047.

Further Reading

Latané, B., & Darley, J. M. (1970). *The unresponsive bystander: Why doesn't he help?* Prentice Hall.

Challenging Convention in Empathy Research: Developing a Mouse Model and Initial Neural Analyses

Jules B. Panksepp, Garet P. Lahvis

Oregon Health and Science University, Portland, OR, United States

"That's not empathy" is a response that we often hear when describing our findings to other scientists. For some of our colleagues, it is far-fetched that individuals from the species *Mus musculus domesticus* could be engaged in something like "empathy." The house mouse, which in some cases has been captive and inbred by biologists for over a century, may serve as a useful "model" system to experiment on the relatively low-hanging fruit of biomedical research, but the concept of empathy goes too far. Empathy, some argue, is a high-level psychological phenomenon reserved for species possessing a highly evolved prefrontal cortex, such as *Homo sapiens*.

Such thinking is associated with a number of tacit premises. First, it implies the possibility that a single region of the brain can underlie a species-specific psychological capacity. It also suggests that empathy arose rapidly in evolution (in the narrow space of primate evolution), an assumption running counter to the generally more graded process of natural selection. Third, scientists often consider the topic of empathy with implicit biases about its fundamental nature. We often talk with each other about "empathy" biased by individualized definitions that are based on our own unique experiences and this leads to considerable confusion.

In this chapter, we will describe a mouse model of empathy along with some of our recent experimental findings. We will provide a context for how this model developed and consider how empathy might be understood as a continuum across species. Several reviews have already provided overviews

Neuronal Correlates of Empathy. http://dx.doi.org/10.1016/B978-0-12-805397-3.00013-9

of the comparative psychology of empathy and include extensive coverage of the relevant studies using rodents (Keum & Shin, 2016; Keysers & Gazzola, 2017; Lahvis, 2016a; Meyza, Bartal, Monfils, Panksepp, & Knapska, 2017; Mogil, 2015; Panksepp & Lahvis, 2011; Panksepp & Panksepp, 2013, 2017; Sivaselvachandran, Acland, Abdallah & Martin, 2016; Wantanabe, 2016). Rather than reviewing these studies again, we will describe our ongoing work with the goal of clarifying the most common misconceptions regarding empathy.

Empathy is a relatively new concept in psychology. As originally defined, it refers to a mental projection that enables an individual to "feel into" a natural environment or an architectural model (Lipps, 1903). In psychology, this term references the moment when an individual assumes an affective state more appropriate to another's situation compared to their own (Hoffman, 2001). In the vernacular, and unfortunately also in the scientific literature, empathy has been conflated with other concepts, such as "sympathy," "compassion," and "prosocial behavior." These terms refer to processes that might be associated with empathy, but are neither synonyms for empathy nor are they processes that empathy requires (Lahvis, 2016a). For instance, sympathy is the ability to understand that another is experiencing distress without necessarily sharing the same emotion. Compassion, or consolation, is associated with a motivation to alleviate another's discomfort. Prosocial behavior results in a benefit to another (i.e., helping), but does not necessarily require empathy (Panksepp & Panksepp, 2017; Silberberg et al., 2014). Thus, these various social phenomena can be supported by a capacity for empathy, but can also be distinct in-and-of themselves. Likewise, sympathy can be fostered by empathy, but it could also result from the ability to adopt another's perspective via cognitive mechanisms (e.g., "Theory of Mind" and perspective taking).

These processes likely vary among species, and across development and environmental contexts. Differences in mating strategies (e.g., monogamy vs. polygamy), infant fostering (e.g., allo-parenting vs. single parent care), and social interest (e.g., social motivation and reward) could also affect such phenotypes. We need to establish specifiable boundaries between empathy and its related psychological processes, phenomena that vary both within and between species. Without clear definitions, we risk unfocused experiments and inaccurate communication about them, particularly as new approaches and techniques emerge.

Perhaps most important is considering the relationship between emotional contagion and empathy. Emotional (or behavioral) contagion is a seemingly reflexive change within the context of an affectively salient event in which an individual spontaneously expresses a behavior resembling the behavior expressed by another individual. Thus, an individual may appear to be expressing discomfort, such as a baby in a nursery crying with other infants, but without implicit knowledge of the source of distress (e.g., is the discomfort internally generated or does it arise from others' experience?).

In this respect, it has been argued that empathy requires the ability for self-recognition, an implicit sense of oneself as an autonomous being. Studies of self-recognition are typically conducted with the mirror self-recognition (MSR) test (Suddendorf & Butler, 2013), which requires a subject that can attend to visual cues on one's face or body. Mice and rats are not usually regarded as possessing "high-level" vision so it is difficult to use the MSR test in these taxa. However, lack of acuity in and/or reliance on a particular sensory modality should not be disqualifying for the ability to recognize self versus other. It would be useful to develop tests of self-recognition that are more appropriate for less visually-attentive rodents, such as tests that rely on the olfactory or auditory senses. Given the lack of such tests, here we will explicitly focus on whether subject responses are expressed in close temporal proximity to an emotional reaction witnessed in a conspecific.

The hypothesis that a laboratory mouse can embody the affective state of a nearby companion requires two capabilities: (1) affect generation (Panksepp, 1998) and (2) communication of such states between individuals (Bishop & Lahvis, 2011). In the remaining part of this chapter, we will discuss our experimental model and highlight where there are similarities or disparities between our findings relative to those of others.

ORIGINS OF THE MODEL AND DEVELOPMENT

In a previous study, we demonstrated that a juvenile mouse exhibits a conditioned place preference for regaining access to their social partners (Panksepp & Lahvis, 2007). Three of the four genetically distinct strains that we evaluated exhibited this response. At one end of the spectrum, mice from the "B6" strain were highly affiliative and expressed a social conditioned place preference (CPP), whereas mice from another strain, "BALB," engaged in less social interaction and did not express a social CPP. We interpreted this as variation in the manifestation of social reward. In humans, empathy is associated with more supportive friendships (Ciarrochi et al., 2016), which are presumably more valued than relationships that are low in empathy. Considering these findings together, we began to wonder whether it was possible to use an experimental framework to ask whether mouse strains that were more responsive to social reward were also more likely to experience empathy. Previous work demonstrated that mice writhed in pain at similar levels when placed next to one another (Langford et al., 2006), but it was unclear if this form of emotional contagion required a psychological experience of empathy because the responses occurred contemporaneously among companions.

To ask how emotions might be shared, we adapted a fear-conditioning paradigm to include a social component. Our "rig" included adjacent chambers where individuals could observe conspecifics undergoing fear conditioning (Fig. 13.1). In accordance with the classical empathy literature, we decided to label the mice within the chambers as "subjects" and referred to the mice undergoing fear conditioning as "objects." We

FIGURE 13.1   **Photograph of the apparatus used for testing empathy in mice.** On the *left* is an observer and on the *right* is a target mouse.

refrained from referring to the fear conditioned mice as "demonstrators" because such terminology might be misleading: It could suggest that an individual is acting intentionally, perhaps attempting to communicate the conditions of the environment or how a particular task is performed. Currently, we have no evidence that fear-conditioned mice behave in such a manner (see later). As our experiments have evolved, however, we realized that the subject-object nomenclature does not fully convey the affective nature of the events that transpire during fear conditioning. Thus, we now interchangeably refer to subject mice in this experimental paradigm as "observers" and object mice as "targets," respectively.

During a typical experiment, an observer is placed inside the fear-conditioning arena, where it becomes familiarized with the environment for a 5-minute period. The observer also receives a single 2–3 s shock (0.5–1.0 mA) halfway through this "habituation" period. The logic for administering a shock to the test mouse before it observes the target is based upon extensive evidence in the empathy literature that suggests prior experience with a distressing stimulus sensitizes (and is perhaps required for) individuals to perceive a similar distress in others (Preston & de Waal, 2002). Though we have not systematically tested this, there is convincing evidence that this may be the case in rodents (Atsak et al., 2011; Rice & Gainer, 1962; Sanders, Mayford, & Jeste, 2013; Sato, Tan, Tate, & Okada, 2015). Importantly, we have never found that this initial experience engenders contextual freezing during the procedures that follow.

Approximately 15 min after habituation, the observers are placed into adjacent compartments, which are 1/4th the size of the conditioning arena, and two target mice are introduced to the conditioning arena. We thought that the presence of multiple targets might amplify signals of social distress and decided to expose observers to two target mice rather than one. In a standard experiment, the target mice and observers have not previously interacted; observers and targets are collected from different home cages and come from different litters that are typically of distinct genetic backgrounds. Thus, targets are socially unfamiliar to observers. We highlight this distinction because there is evidence that social familiarity can modulate empathy-related behaviors in rodents (Ben-Ami Bartal, Rodgers, Bernardez Sarria, Decety, & Mason, 2014; Gonzalez-Liencres, Juckel, Tas, Friebe, & Brüne, 2014; Jeon et al., 2010; Jones, Riha, Gore, & Monfils, 2014; Kiyokawa, Honda, Takeuchi, & Mori, 2014; Langford et al., 2006; Li et al., 2014).

Targets undergo a standard fear-conditioning procedure, which entails presentation of a 30-second tone (1 khz, 85 db) that co-terminates with the same level of shock delivered to the observer during the habituation period. These experiences are spaced by 90-second intervals and are repeated 10 times. After each shock is administered, observers from the B6 genetic background orient toward the conditioning arena and briefly close their eyes, but they do not freeze. Target mice exhibit an approximate 4% increase in heart rate (≈increase of 30 beats/minute) relative to baseline, which peaks following the administration of the first shock. Observers increase heart rate to a much smaller degree (1%) across the same time frame, which is followed by a progressive slowing to ≈5% below baseline by the end of conditioning. By contrast, the heart rate of targets returns to a baseline-level by the end of conditioning. Relative to observers, target mice also display hyperglycemia after conditioning, which may indicate differential activation of the hypothalamic-pituitary axis.

This "conditioning" procedure is repeated on the next day and testing begins 15 min later. An observer is reintroduced to the conditioning arena where it is exposed to the same tones that were presented during conditioning. Observer mice from the B6 genetic background freeze for approximately 10%–20% (≈3–6 s) of the time elapsed during tone presentation, although there is substantial variation (see later). This response is specific to experiencing the combination of distressed target mice and the tone, as unpaired or isolated presentation of these stimuli has a less robust influence on freezing (Fig. 13.2).

When observers are derived from the BALB genetic background, a different picture emerges (Chen, Panksepp, & Lahvis, 2009). Like B6, BALB observers orient their head toward the conditioning arena, close their eyes and do not freeze after shock delivery to targets. However, BALB observers do not express the heart rate slowing that is expressed by B6 observers. Moreover, when placed back into the conditioning arena, BALB mice express minimal levels of freezing to presentation of the tone (freezing behavior is actually often completely undetectable). Thus, whereas

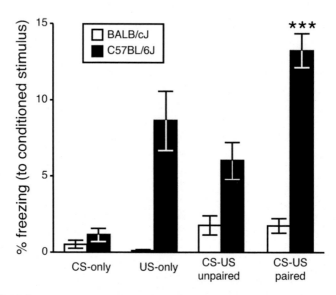

FIGURE 13.2 **Freezing responses of observer mice to tones (conditioned stimulus, CS) presented in the conditioning after experiencing two target mice that were exposed to various combinations of the CS and shock (unconditioned stimulus, US).** Data are presented as mean ± std. error. N's = 14–26 mice per genotype/condition. Statistical significance was determined via Tukey's honestly significantly different test(s) following a 2 × 4 analysis of variance.

the immediate behavioral responses of both mouse strains to the distress signals of a conspecific are similar, their subsequent physiological and behavioral responses differ. Decreased heart rate (presumably via parasympathetic activation) and freezing to a tone predictive of others' distress are thus indicative of the B6 empathy response.

Compared to the observer responses described above that were recorded 15 min after conditioning, observers expressed more robust freezing responses to conspecific distress-associated tones if they were tested 24 h after they experienced target mouse conditioning (Panksepp & Lahvis, 2016). However, if observer mice were raised in social isolation during adolescence and prior to conditioning/testing, they fail to express an increased freezing response. Rather, the freezing response of isolated B6 observer mice diminishes over the period separating the 15 min- and 24 h-time points. Interactions with peers during adolescent development therefore support expression of this vicarious fear phenotype whereas social deprivation appears to be deleterious.

These initial findings suggested a positive relationship between affiliative behavior and empathy. However, when we more recently expanded these studies to include 4 additional mouse strains (BTBR, DBA, FVB, and Swiss Webster) along with B6 and BALB, we found no evidence for an association between genetically based differences in affiliative behavior and empathy (Panksepp, 2015), which suggests independence between gregariousness and the expression of empathy. One possibility is that a capacity for empathy in our experimental model is most robust under "intermediate" levels of social motivation. Whereas low social engagement may render individuals less attentive to the social cues associated with distress, very high levels of social motivation could bias individuals toward excessive social reward seeking, perhaps distracting them from attending to the affective state of conspecifics.

THE SOUND OF DISCOMFORT

When target mice experience a shock, they vocalize at frequencies audible to the human ear. These vocalizations are emitted at a rate of 8–12 per shock and have a fundamental frequency of 8–19 khz. Given the recent plethora of studies of rodent auditory communication (for reviews, see Barker, Simmons, & West, 2015; Brudzynski, 2015; Simola, 2015), it is important to highlight that these vocalizations are often mistaken for ultrasonic vocalizations (USVs). Mice emit USVs almost exclusively during social contexts or after they are administered drugs of abuse, and we have rarely observed USVs during any phase of fear conditioning, irrespective of the presence of observer mice. By contrast, the audible vocalizations mice emit when they are shocked occur reliably when the stimulus is delivered, independent of the presence of observer mice. Unlike USVs, which are emitted as narrow bandwidth (i.e., high energy concentration), frequency modulated calls (Lahvis, Alleva, & Scattoni, 2011), these vocalizations have a broadband spectrographic profile (Fig. 13.3A). These squeaks are also emitted when mice receive an injection, an ear notch, or are in the presence of an aggressive conspecific. The amplitude and frequency of occurrence of these mouse vocalizations appears to vary with the intensity of the shock (Fig. 13.3A).

Since observers and targets are separated by a wall of metal bars, and conditioned/tested under dim red or infrared light, we know that the distress of target mice is not communicated by visual or somatosensory cues. To determine whether target mouse vocalizations communicate distress, we exposed observer mice to tones that co-terminated with auditory playbacks of mice undergoing shock. All other aspects of the experiment resembled those described earlier. Within this context, B6 observers subsequently froze to tones that had been paired with these distress vocalizations, but BALB observers did not (Chen et al., 2009). The B6-BALB difference in vicarious fear can thus be reproduced when the presence of target mice is replaced with vocalizations played through a speaker. Importantly, at the age when testing occurs (5–8 weeks), there is no

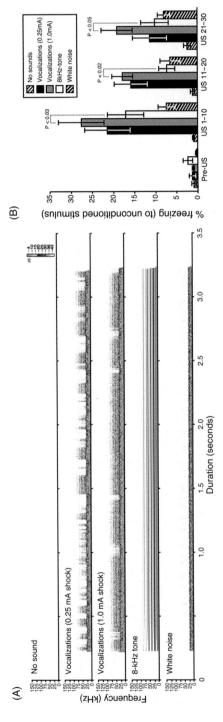

FIGURE 13.3　**Sonograms of unconditioned stimuli and behavioral responses to these stimuli.** (A) For sonograms 2 and 3, sounds were recorded from a pair of Swiss Webster mice (one per sex). 8-kHz pure tones and white noise were generated with a software program from Cleversys Inc. (B) After 5 days of habituation, observer mice were presented with one of these isolated auditory stimuli (3 s "on" duration, 27 s "off" duration, 30 cycles). Percentage of freezing to each stimulus is presented on the ordinate. Data are presented as mean ± std. error. N's = 8–16 mice per group. Statistical significance was determined via Tukey's honestly significantly different test(s) following a 5-way analysis of variance with trial as a repeated measure.

evidence that B6 or BALB mice exhibit differences in processing sounds of the same fundamental frequency, as measured by the auditory brainstem response (Zheng, Johnson, & Erway, 1998), indicating that this difference in vicarious fear is not likely related to a sensory processing deficit.

Using classical conditioning terminology for a typical Pavlovian scenario, the tone would be referred to as a conditioned stimulus (CS) and the shock as an unconditioned stimulus (US). Through repeated pairings, a target mouse learns that the CS is predictive of discomfort engendered by the US. As described earlier, for an observer mouse the scenario is somewhat different; the US in this case appears to be the vocalizations of target mice. Observers adopt the distress of target mice via the sound of discomfort. Although we cannot rule out a role for olfactory signaling of fear, our experiments show that auditory cues alone are sufficient to communicate this affective experience.

NEURAL CORRELATES OF VICARIOUS FEAR

The radionuclide ^{18}F-fluordeoxyglucose (FDG) is the prototypical tracer used in positron emission tomography (PET) studies of brain glucose metabolism. FDG can be used to identify active regions of the brain of a rodent when it is directly experiencing the US (i.e., what target mice experience) or witnessing the distress of conspecifics (i.e., what observer mice experience). We developed a technique where, unlike PET, subjects do not need to be food deprived or restrained, and the tracer can be absorbed via the peritoneum (Fig. 13.4A). Following direct or vicarious fear expe-

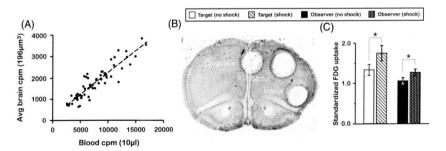

FIGURE 13.4 **Measuring glucose metabolism in the B6 mouse brain.** (A) Following an i.p. injection of 400 microCuries FDG, concentration in the blood was a strong predictor of the amount that reaches the brain. (B) Regions of interest are biopsied from 250-micron "slabs" and -511 kEV photons are quantified in a well-type gamma counter. (C) Differences in FDG uptake in B6 mouse premotor cortex. Uptake values were standardized: (cpm [brain]/ cpm [blood]) (blood sugar/blood sugar [average]) (1/weight) (100). Data are presented as mean ± std. error. N's = 12–22 mice per group. Statistical significance was determined via Tukey's honestly significantly different test(s) following a 2 × 2 analysis of variance.

riences, we then utilized the Palkovits punch to biopsy brain regions of interest (Fig. 13.4B) and samples were assessed for changes in glucose metabolism in a well-type gamma counter.

In an initial study, we habituated subjects to injection and the respective environment for 2 days. Then, B6 mice were either exposed directly to shocks (2-second duration, 1.0 mA intensity, 10 cycles with 128-second inter-stimulus interval) or to target mice receiving the same shocks. More than 50 brain regions from areas thought to be involved in empathy and social behavior were collected. We also assessed control regions (e.g., visual cortex). Our overall goal was to identify how the brain responds to the unexpected experience of conspecific distress.

In this experiment, brain areas could be categorized as responsive to (1) directly experienced fear, (2) vicariously experienced fear, or (3) to both conditions. We found that target mice exhibited increased metabolic responses in frontal association cortex, nucleus accumbens septi, and a wide extent of the insular cortex, cingulate cortex, and amygdala, as well as posterior limb areas of the somatosensory cortex, the periaquiductal grey, and a region encompassing cerebellar lobules 3/4/5. Regions activated in both observer and target mice included frontal regions of the cingulate, somatosensory, and motor (Fig. 13.4C) cortex, as well as anterior aspects of the medial thalamic nuclei. Observer mice also exhibited activation in some distinct brain areas, including the auditory cortex, a medial aspect of the insular cortex, and an amygdalar region composed of posterior basal and cortical nuclei.

These observations provide insight into the neural circuitry that may underlie the experience of vicarious fear and complement other rodent empathy models, utilizing inducible transcription factor activity and electrophysiological approaches (Borg et al., 2017; Jeon et al., 2010; Knapska et al., 2006; Meyza et al., 2015). Collectively, these studies pave the way for a comprehensive evaluation of a possible "mirroring" system in rodent brains. Genetic approaches to monitoring neuronal activation in vivo (Guenthner, Miyamichi, Yang, Heller, & Luo, 2013; Reijmers & Mayford, 2009) might allow us to determine whether neurons that monitor self-distress versus vicarious distress are unique populations or the same cells (e.g., "mirror neurons").

RELATION TO BIOMEDICAL MODELS

Autism spectrum disorders (ASDs), depression and posttraumatic stress disorder share various levels of disruption in responding to another's emotional state or in accessing memories imbued by socioemotional content. In one clinical assessment of autism, a clinician "accidentally" drops a stack of papers or hits his/her finger with a mallet, and then expresses vocal and gestural expressions of surprise or pain, respectively. Typically developing children visually orient to the object and the face of the clinician, and then express a congruent emotional reaction and may respond prosocially, whereas some individuals with ASD may not (Charman et al., 1997; Dawson et al., 2004). The mouse model described here bears some face validity to this clinical assessment and such preclinical models could be revolutionary for screening/developing pharmacological compounds for therapy.

In a preliminary study, we attempted to stimulate the central oxytocin system of B6 observers as they witnessed the distress of target mice. Oxytocin and its paralogue neuropeptides have gained a considerable amount of attention in recent years for their role in social processes, particularly parental behavior and social bonding (Feldman, 2016; Johnson & Young, 2015). Indeed, as this chapter was being written there were 21 active clinical studies involving "oxytocin" administration and "empathy" (ClinicalTrials.gov) in the United States. We had thus hypothesized that pharmacological manipulation of oxytocin receptors might facilitate vicarious fear responding.

B6 mice were given an intraperitoneal injection of WAY-267464 15 min prior to observing target distress paired with tone presentation (as described earlier). WAY-267464 is a small-molecule, blood-brain barrier permeable oxytocin receptor agonist that decreases anxiety in mice at low doses (Ring et al., 2010) and increases social behavior in rats at high doses (Hicks et al., 2012). We employed a within-subject experimental design, administering 0, 1, 10, or 100 mg/kg WAY-267464 versus vehicle in control groups (days 1, 3, 5, and 7 respectively). Observers were tested for tone-induced freezing 15 min after experiencing target distress and again 24 h post-conditioning (days 2, 4, 6, and 8).

Surprisingly, no differences in tone-induced freezing between WAY-267464 treated and control individuals were found across the doses, or the short-term (Fig. 13.5) and long-term time points. The most prominent difference was a steep decline in tone-induced freezing across experimental days (both 15 min and 24 h postconditioning). This decrease was despite repeated pairings of the tone with conspecific distress on days 1, 3, 5, and 7, eliminating the possibility that the decline was due to "extinction." Rather it appears that observer mice are learning something new. Since observer and target mice live in different cages, mice do not have the chance to interact with each other post-distress, and it is possible that "learned safety" occurs (Kong, Monje, Hirsch, & Pollak, 2014). It is also conceivable that observers habituate to the vocalizations of target mice across repeated experiences, similar to "the boy who cried wolf."

To avoid high-dose exposure potentially affecting subsequent responding, we utilized a procedure that included increasing concentrations across the experiment, with the highest dose of WAY-267464 administered on final day of conditioning. Thus, we inadvertently opposed ascending dosages against a behavioral response that decreases across repeated

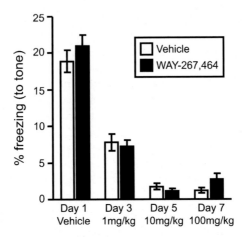

FIGURE 13.5   **Responses of B6 observer mice to social distress-associated tones across days and following administration of the nonpeptide oxytocin receptor agonist WAY-267464.** Administration of the agonist did not affect vicarious fear 15 min after the experience with conspecific distress. N's = 8–14 mice per treatment group. Data were analyzed with a 2-way analysis of variance with day as a repeated measure.

testing. This complicates making strong interpretations about oxytocin's role in mouse empathy. Now that we know tone-induced freezing diminishes across days, a high dose during the first day of the procedure may yield different results. Additionally, the level of vicarious freezing by B6 observers may represent a "ceiling" response. Oxytocin receptor stimulation could therefore be more efficacious in strains such as BALB, which show much lower levels of acquired vicarious fear relative to B6. It should also be noted that oxytocin was recently shown to facilitate consolation behavior in a prairie vole model (Burkett et al., 2016). Thus, it is possible that oxytocin is important for providing social comfort following distress, but not for empathy per se.

CONTAGION OR EMPATHY?

As described earlier, there are clear definitional and conceptual boundaries that dissociate emotional contagion from empathy. There nevertheless remains a murky usage of such terms in the literature. In many cases, contagion appears to be a relatively reflexive behavioral process, including responses such as yawning (Moyaho, Rivas-Zamudio, Ugarte, Eguibar, & Valencia, 2015) and scratching (Yu, Barry, Hao, Liu, & Chen, 2017). However, other examples of contagion, including crying and writhing in pain, clearly involve affective arousal.

In addition to the concept of "self-recognition" (see earlier), temporal contiguity and a similar magnitude of target-observer response, parity of physiological outcomes, and whether an observer experiences self-distress have all been proposed as factors that can distinguish contagion from empathy. We do not yet know whether observer mice find the procedures described in this chapter aversive, but CPP tests are invaluable for determining the degree to which an animal finds a situation emotionally salient and are warranted in future studies.

In a recent study, we exposed observers to the vocalizations of target mice receiving fear conditioning and compared them to control sounds, such as white noise or pure tones at the fundamental frequency of these vocalizations. In contrast to the studies described above, we administered stimuli for a longer duration (3 s) and a faster rate (every 27 s), resulting in a more extreme experience for observers (i.e., 30 total compared to 10 shorter and lower-intensity shocks provided in the Chen et al. study). In this context, B6 observers displayed contagious freezing when listening to the vocalizations of target mice relative to control sounds (Fig. 13.3B).

Consistent with previous behavioral studies (Jeon et al., 2010; Gonzalez-Liencres et al., 2014), these findings suggest that the same individual can express emotional contagion or empathy, and that the particular form of expression is heavily influenced by the severity of the environmental context. Future studies of observer's brain responses across varying levels of conspecific distress will help delineate the extent to which contagion and empathy share overlapping or distinctive neural substrates. Insofar that empathy requires self-recognition, as described above, such studies might also help to elucidate where in the brain a sense of self is represented.

CONCLUSION

Our studies of empathy began nearly a decade ago and we now know that many of the findings in mice go hand in hand with what is known about empathy in primate species, including humans. However, with the passage of time and perhaps some distance from the experiments, we also see several problems with this experimental approach. What should be most obvious to anyone reading this chapter is that we shock our mice to engender an affective experience. Empathy is elicited by the emotion of a conspecific and we felt that such momentary pain was a more compassionate approach to generating pain than the long-term discomfort of formalin injections into the paw or acetic acid injections into the peritoneum. Nonetheless, such experiences are neither subtle nor natural for the mouse and we must consider the possibility that more pleasurable experiences with conspecifics might engender empathic responses as well (Panksepp & Lahvis, 2011; Panksepp & Panksepp, 2017).

Also important in hindsight is that these mice were housed under very unnatural conditions. The house mouse has a natural home range that is hundreds of thousands fold larger than the small laboratory cage it is afforded (Lahvis, 2017a,b) and because this highly restricted and impoverished form of captivity offers no agency or even simple options for social refuge, it can powerfully modify brain development (Lahvis, 2016b). We need to consider these caveats when we interpret our data and as we conceive of future experiments.

References

Atsak, P., Orre, M., Bakker, P., Cerliani, L., Roozendaal, B., Gazzola, V., et al. (2011). Experience modulates vicarious freezing in rats: A model for empathy. *Public Library of Science One, 6*, e21855. doi: 10.1371/journal.pone.0021855.

Barker, D. J., Simmons, S. J., & West, M. O. (2015). Ultrasonic vocalizations as a measure of affect in preclinical models of drug abuse: A review of current findings. *Current Neuropharmacology, 13*, 193–210. doi: 10.2174/1570159X13999150318113642.

Ben-Ami Bartal, I. B., Rodgers, D. A., Bernardez Sarria, M. S., Decety, J., & Mason, P. (2014). Pro-social behavior in rats is modulated by social experience. *Elife, 3*, e01385. doi: 10.7554/eLife.01385.

Bishop, S. L., & Lahvis, G. P. (2011). The autism diagnosis in translation: Shared affect in children and mouse models of ASD. *Autism Research, 4*, 317–335. doi: 10.1002/aur.216.

Borg, J.S., Srivastava, S., Lin, L., Heffner, J., Dunson, D., Dzirasa, K., & de Lecea, L. (2017). Rat intersubjective decisions are encoded by frequency-specific oscillatory contexts. Brain and Behavior, 7, e00710. doi: 10.1002/brb3.710.

Brudzynski, S. M. (2015). Pharmacology of ultrasonic vocalizations in adult rats: Significance, call classification and neural substrate. *Current Neuropharmacology, 13*, 180–192. doi: 10.2174/1570159X13999150210141444.

Burkett, J. P., Andari, E., Johnson, Z. V., Curry, D. C., de Waal, F. B., & Young, L. J. (2016). Oxytocin-dependent consolation behavior in rodents. *Science, 351*, 375–378. doi: 10.1126/science.aac4785.

Charman, T., Swettenham, J., Baron-Cohen, S., Cox, A., Baird, G., & Drew, A. (1997). Infants with autism: An investigation of empathy, pretend play, joint attention, and imitation. *Developmental Psychology, 33*, 781–789. doi: 10.1037/0012-1649.33.5.781.

Chen, Q., Panksepp, J. B., & Lahvis, G. P. (2009). Empathy is moderated by genetic background in mice. *Public Library of Science One, 4*, e4387. doi: 10.1371/journal.pone.0004387.

Ciarrochi, J., Parker, P. D., Sahdra, B. K., Kashdan, T. B., Kiuru, N., & Conigrave, J. (2016). When empathy matters: The role of sex and empathy in close friendships. *Journal of Personality*. doi: 10.1111/jopy.12255 [Epub ahead of print].

Dawson, G., Toth, K., Abbott, R., Osterling, J., Munson, J., Estes, A., et al. (2004). Early social attention impairments in autism: Social orienting, joint attention, and attention to distress. *Developmental Psychology, 40*, 271–283. doi: 10.1037/0012-1649.40.2.271.

Feldman, R. (2016). The neurobiology of mammalian parenting and the biosocial. context of human caregiving. *Hormones and Behavior, 77*, 3–17. doi: 10.1016/j.yhbeh.2015.10.001.

Gonzalez-Liencres, C., Juckel, G., Tas, C., Friebe, A., & Brüne, M. (2014). Emotional contagion in mice: The role of familiarity. *Behavioural Brain Research, 263*, 16–21. doi: 10.1016/j.bbr.2014.01.020.

Guenthner, C. J., Miyamichi, K., Yang, H. H., Heller, H. C., & Luo, L. (2013). Permanent genetic access to transiently active neurons via TRAP: Targeted recombination in active populations. *Neuron, 78*, 773–784. doi: 10.1016/j.neuron.2013.03.025.

Hicks, C., Jorgensen, W., Brown, C., Fardell, J., Koehbach, J., Gruber, C. W., et al. (2012). The nonpeptide oxytocin receptor agonist WAY 267, 464: Receptor-binding profile, prosocial effects and distribution of c-Fos expression in adolescent rats. *Journal of Neuroendocrinology, 24*, 1012–1029. doi: 10.1111/j.1365-2826.2012.02311.x.

Hoffman, M. (2001). *Empathy and moral development*. Cambridge: Cambridge University Press.

Jeon, D., Kim, S., Chetana, M., Jo, D., Ruley, H. E., Lin, S. Y., et al. (2010). Observational fear learning involves affective pain system and Cav1.2 Ca2+ channels in ACC. *Nature Neuroscience, 13*, 482–488. doi: 10.1038/nn.2504.

Johnson, Z. V., & Young, L. J. (2015). Neurobiological mechanisms of social attachment and pair bonding. *Current Opinion in Behavioral Sciences, 3*, 38–44. doi: 10.1016/j.cobeha.2015.01.009.

Jones, C. E., Riha, P. D., Gore, A. C., & Monfils, M. H. (2014). Social transmission of Pavlovian fear: Fear-conditioning by-proxy in related female rats. *Animal Cognition, 17*, 827–834. doi: 10.1007/s10071-013-0711-2.

Keum, S., & Shin, H. S. (2016). Rodent models for studying empathy. *Neurobiology of Learning and Memory, 135*, 22–26. doi: 10.1016/j.nlm.2016.07.022.

Keysers, C., & Gazzola, V. (2017). A plea for cross-species social neuroscience. *Current Topics in Behavioral Neuroscience, 30*, 179–191. doi: 10.1007/7854_2016_439.

Kiyokawa, Y., Honda, A., Takeuchi, Y., & Mori, Y. (2014). A familiar conspecific is more effective than an unfamiliar conspecific for social buffering of conditioned fear responses in male rats. *Behavioural Brain Research, 267*, 189–193. doi: 10.1016/j.bbr.2014.03.043.

Knapska, E., Nikolaev, E., Boguszewski, P., Walasek, G., Blaszczyk, J., Kaczmarek, L., et al. (2006). Between-subject transfer of emotional information evokes specific pattern of amygdala activation. *Proceedings of the National Academy of Sciences (USA), 103*, 3858–3862. doi: 10.1073/pnas.0511302103.

Kong, E., Monje, F. J., Hirsch, J., & Pollak, D. D. (2014). Learning not to fear: Neural correlates of learned safety. *Neuropsychopharmacology, 39*, 515–527. doi: 10.1038/npp.2013.191.

Lahvis, G. P. (2016a). Social reward and empathy as proximal contributions to altruism: The camaraderie effect. In M. Wöhr, & S. Krach (Eds.), *Social behavior from rodents to humans* (pp. 127–157). Switzerland: Springer International Publishing.

Lahvis, G. P. (2016b). Rodent models of autism, epigenetics, and the inescapable problem of animal constraint. In J. C. Gewirtz, & Y. K. Kim (Eds.), *Animal models of behavior genetics* (pp. 265-301). New York: Springer Nature.

Lahvis, G. P. (2017a). Animal welfare: Make animal models more meaningful. *Nature, 543*, 623–623. doi: 10.1038/543623d.

Lahvis, G. P. (2017b). Unbridle biomedical research from the laboratory cage. *eLife, 6*, e27438. doi: 10.7554/eLife.27438.

Lahvis, G. P., Alleva, E., & Scattoni, M. L. (2011). Translating mouse vocalizations: Prosody and frequency modulation. *Genes, Brain and Behavior, 10*, 4–16. doi: 10.1111/j.1601-183X.2010.00603.x.

Langford, D. J., Crager, S. E., Shehzad, Z., Smith, S. B., Sotocinal, S. G., Levenstadt, J. S., et al. (2006). Social modulation of pain as evidence for empathy in mice. *Science, 312*, 1967–1970. doi: 10.1126/science.1128322.

Li, Z., Lu, Y. F., Li, C. L., Wang, Y., Sun, W., He, T., et al. (2014). Social interaction with a cagemate in pain facilitates subsequent spinal nociception via activation of the medial prefrontal cortex in rats. *Pain, 155*, 1253–1261. doi: 10.1016/j.pain.2014.03.019.

Lipps, T. (1903). Einfühlung, innere Nachahmung, und Organepfindungen. *Archiv für die gesamte Psychologie, 3*, 185–204.

Meyza, K., Bartal, I. B., Monfils, M. H., Panksepp, J. B., & Knapska, E. (2017). The biological roots of empathy: Through the lens of rodent models. *Neuroscience and Biobehavioral Reviews, 76*, 216–234. doi: 10.1016/j.neubiorev.2016.10.028.

Meyza, K., Nikolaev, T., Kondrakiewicz, K., Blanchard, D. C., Blanchard, R. J., & Knapska, E. (2015). Neuronal correlates of asocial behavior in a BTBR T (+) Itpr3(tf)/J mouse model of autism 1. *Frontiers in Behavioral Neuroscience, 9*, 199. doi: 10.3389/fnbeh.2015.00199.

Mogil, J. S. (2015). Social modulation of and by pain in humans and rodents. *Pain, 156*, 35–41. doi: 10.1097/01.j.pain.0000460341.62094.77.

Moyaho, A., Rivas-Zamudio, X., Ugarte, A., Eguibar, J. R., & Valencia, J. (2015). Smell facilitates auditory contagious yawning in stranger rats. *Animal Cognition, 18,* 279–290. doi: 10.1007/s10071-014-0798-0.

Panksepp, J. (1998). *Affective neuroscience: the foundation of human and animal emotions.* New York: Oxford University Press.

Panksepp, J., & Panksepp, J. B. (2013). Toward a cross-species understanding of empathy. *Trends in Neurosciences, 36,* 489–496. doi: 10.1016/j.tins.2013.04.009.

Panksepp, J. B (2015). Modeling vicarious fear in adolescent mice. In *Symposium talk presented at the 24th annual meeting of the International Behavioral Neuroscience Society* Victoria, BC, Canada June.

Panksepp, J. B., & Lahvis, G. P. (2007). Social reward among juvenile mice. *Genes, Brain & Behavior, 6,* 661–671. doi: 10.1111/j.1601-183X. 2006.00295.x.

Panksepp, J. B., & Lahvis, G. P. (2011). Rodent empathy and affective neuroscience. *Neuroscience and Biobehavioral Reviews, 35,* 1864–1875. doi: 10.1016/j.neubiorev.2011.05.013.

Panksepp, J. B., & Lahvis, G. P. (2016). Differential influence of social versus isolate housing on vicarious fear in adolescent mice. *Behavioral Neuroscience, 130,* 206–211. doi. org/10.1037/bne0000133.

Panksepp, J. B., & Panksepp, J. (2017). Empathy through the ages: a comparative perspective on rodent models of shared emotion. In J. Call (Ed.), *Handbook of comparative psychology* (pp. 765–792). Washington DC: American Psychological Association.

Preston, S. D., & de Waal, F. B. M. (2002). Empathy: Its ultimate and proximate bases. *Behavioral and Brain Sciences, 25,* 1–20. doi: 10.1017/s0140525x02000018.

Reijmers, L., & Mayford, M. (2009). Genetic control of active neural circuits. *Frontiers in Molecular Neuroscience, 2,* 27. doi: 10.3389/neuro.02.027.2009.

Rice, G. E., & Gainer, P. (1962). "Altruism" in the albino rat. *Journal of Comparative Physiology and Psychology, 55,* 123–125. doi: 10.1037/h0042276.

Ring, R. H., Schechter, L. E., Leonard, S. K., Dwyer, J. M., Platt, B. J., Graf, R., et al. (2010). Receptor and behavioral pharmacology of WAY-267464, a non-peptide oxytocin receptor agonist. *Neuropharmacology, 58,* 69–77. doi: 10.1016/j.neuropharm.2009.07.016.

Sanders, J., Mayford, M., & Jeste, D. (2013). Empathic fear responses in mice are triggered by recognition of a shared experience. *Public Library of Science One, 8,* e74609. doi: 10.1371/journal.pone.0074609.

Sato, N., Tan, L., Tate, K., & Okada, M. (2015). Rats demonstrate helping behavior toward a soaked conspecific. *Animal Cognition, 18,* 1039–1047. doi: 10.1007/s10071-015-0872-2.

Silberberg, A., Allouch, C., Sandfort, S., Kearns, D., Karpel, H., & Slotnick, B. (2014). Desire for social contact, not empathy, may explain "rescue" behavior in rats. *Animal Cognition, 17,* 609–618. doi: 10.1007/s10071-013-0692-1.

Simola, N. (2015). Rat ultrasonic vocalizations and behavioral neuropharmacology: From the screening of drugs to the study of disease. *Current Neuropharmacology, 13,* 164–179. doi: 10.2174/1570159X13999150318113800.

Sivaselvachandran, S., Acland, E. L., Abdallah, S., & Martin, L. J. (2016). Behavioral and mechanistic insight into rodent empathy. *Neuroscience and Biobehavioral Reviews.* doi: 10.1016/j.neubiorev.2016.06.007. [Epub ahead of print]

Suddendorf, T., & Butler, D. L. (2013). The nature of visual self-recognition. *Trends in Cognitive Science, 17,* 121–127. doi: 10.1016/j.tics.2013.01.004.

Wantanabe, S. (2016). Evolutionary origins of empathy. In D. F. Watt, & J. Panksepp (Eds.), *Psychology and neurobiology of empathy.* New York: Nova Science Publishers, Inc.

Yu, Y. Q., Barry, D. M., Hao, Y., Liu, X. T., & Chen, Z. F. (2017). Molecular and neural basis of contagious itch behavior in mice. *Science, 355,* 1072–1076. doi: 10.1126/science.aak9748.

Zheng, Q. Y., Johnson, K. R., & Erway, L. C. (1998). Assessment of hearing in 80 inbred strains of mice by ABR threshold analyses. *Hearing Research, 130,* 94–107. doi: 10.1016/S0378-5955(99)00003-9.

Lack of Empathy—Mouse Models

Ksenia Z. Meyza

Nencki Institute of Experimental Biology, Polish Academy of Sciences, Warsaw, Poland

EMPATHY DEFICITS IN HUMANS

Human empathy is a complex phenomenon. Most researchers describe it as the ability to perceive and vicariously experience the emotional states of others, leading to an initiation of prosocial behaviors (for review see Decety, Bartal, Uzefovsky, & Knafo-Noam, 2016). Deficits in empathy arise from either (or both) of two factors: (1) the inability to understand the emotional state of other people and/or (2) the inability to be affected by it. Smith (2006) proposed a model dividing empathy impairments into functional classes to reflect that diversity. It includes four distinct categories: (1) Cognitive Empathy Deficit Disorder, (2) Emotional Empathy Deficit Disorder (EEDD), (3) General Empathy Deficit Disorder (GEDD), and (4) General Empathy Surfeit Disorder (GESD). Cognitive empathy, often regarded synonymous with the Theory of Mind, is defined here as the capability to understand another's perspective or mental state, while emotional empathy entails the ability to respond with an appropriate emotion to emotions and mental states of others.

The first of these classes, CEDD is characterized by deficient understanding of the emotional states of others and intact emotional empathy. These features are, according to Smith, typical of neurodevelopmental disorders such as the autism spectrum disorder (ASD). Smith (2009) further postulated that the emotional empathy might even be increased in ASD patients, although this notion has met with considerable opposition (Minio-Paluello, Lombardo, Chakrabarti, Wheelwright, & Baron-Cohen, 2009).

Neuronal Correlates of Empathy. http://dx.doi.org/10.1016/B978-0-12-805397-3.00014-0

The deficiency in understanding of emotions was found to be prominent in ASD patients, both in the case of emotions of others and in identification of emotions in oneself (alexithymia). Structural magnetic resonance data suggest that alexithymia and ASD result from altered function of distinct neuronal circuits (Bernhardt et al., 2014). In line with this, identification of certain (especially emotionally charged) facial expressions was found to be uniformly impaired in ASD patients (Crawford, Moss, Anderson, Oliver, & McCleery, 2015; Walsh, Creighton, & Rutherford, 2016), while the degree of alexithymia was found to modulate the activation of empathy-relevant brain structures similarly in control and ASD subjects (Bird et al., 2010; Silani et al., 2008). The co-occurrence of these two (~50% of ASD subjects display severe levels of alexithymia, Berthoz and Hill, 2005), however, may exacerbate emotion recognition deficits in the ASD population (Cook, Brewer, Shah, & Bird, 2013).

Altered expression of emotions by ASD individuals is another source of complication affecting their empathic responses. Brewer and coworkers (2016) showed that when ASD patients are used as stimuli in facial expression recognition tasks, their expressions are less well recognized both by typically developing (TD) individuals and ASD individuals.

Individuals with ASD self-report diminished empathic concern but higher or intact personal distress upon exposure to a vicarious pain experience (Dziobek et al., 2008; Minio-Paluello, Baron-Cohen, Avenanti, Walsh, & Aglioti, 2009). While such reaction to vicarious pain might be an indication of increased emotional empathy, it refers instead to the discomfort experienced during a novel, possibly difficult to understand, social situation. Simpler forms of empathic responses, including facial mimicry (Press, Richardson, & Bird, 2010) and contagious yawning, initially considered absent in ASD subjects were recently demonstrated to be present in the affected population (Usui et al., 2013).

On the neuronal level, ASD patients display aberrant activation within the middle cingulate and ventromedial prefrontal cortices, as well as anterior insula in self-inferred tasks (Lombardo, Chakrabarti, & Bullmore, 2010; Pfeifer et al., 2013) but respond with intact activation of these structures while referring to others with ASD (Komeda et al., 2015). On the other hand, their response to TD controls in the latter task was decreased. The TD controls showed intact activation to other TD subjects and a decreased response to ASD subjects (Komeda et al., 2015). A first-hand pain experience in ASD patients results in activation of similar brain circuits to that of TD controls (Krach et al., 2015). The response to vicarious experience of pain was also recently found to be intact (Hadjikhani et al., 2014), despite earlier reports showing decreased neurophysiological responsiveness in subjects with Asperger's syndrome (Minio-Paluello and Baron-Cohen et al., 2009).

The second class of impairment, EEDD entails full awareness of the emotional states of others combined with a complete lack of ability to share these emotions. Smith (2006) suggested that people with EEDD would have excellent social skills allowing them to manipulate and deceive others. Their intact cognitive empathy would let them appear as if they share other people's concerns by sheer prediction of what the socially acceptable response should be. These traits are characteristic of psychopathy, conduct disorder and antisocial personality disorder (Blair, 2013; Lockwood et al., 2013). Several studies have compared behavioral and neuronal responses in CEDD and EEDD (for review see Lockwood, 2016) and found that, unlike ASD patients, children, and adults with psychopathic traits display reduced reactivity to distress in others and show diminished affective empathic responses. Adolescents with psychopathic traits displayed reduced activity in brain structures related to empathy (Marsh et al., 2013; Michalska, Zeffiro, & Decety, 2016) especially in response to pain inflicted on other people.

Tests performed on adult, incarcerated offenders with different degrees of psychopathy showed that individuals scoring high on psychopathy trait, despite intact response of empathy-related brain structures to first-hand pain experience, exhibited significantly less activation in the ventromedial prefrontal cortex, lateral orbitofrontal cortex, and periaqueductal gray, but intact or increased activation in other areas (e.g., insula), while responding to facial expressions of pain (Decety, Skelly, & Kiehl, 2013). Observing hands exposed to harmful treatment resulted with hypoactivation of the amygdala of the psychopathic offenders, but upon being instructed to empathize with the person depicted in the video, the difference between psychopaths and nonpsychopathic controls was reduced (Meffert, Gazzola, den Boer, Bartels, & Keysers, 2013).

The third category of impairment, GEDD encompasses lack of both forms of empathy, leaving the affected person both unable to understand and share the emotional cues coming from other people. At the same time, such people would not be able to follow social convention, which would cause them to withdraw from society. Similar characteristics were reported for schizophrenia patients, who tend to isolate themselves (Derntl et al., 2015; Green, Horan, & Lee, 2015). A recent study investigating neuronal response to viewing videos of painful medical treatments showed that schizophrenics shared activation of empathy-relevant brain structures with the control group. Higher activation, however, was observed when patients were asked to imagine the procedure being inflicted on others versus themselves (Horan et al., 2016). The control group in that study displayed a reverse effect. Emotional deficits of subjects with schizophrenia, including the decreased ability to recognize emotion in others, were also associated with decreased activity in the cerebellum (for a review see Mothersill, Knee-Zaska, & Donohoe, 2016).

The GESD impairment represents a phenotype opposite to that of GEDD. Hypersociability, with enhanced empathic concern and affect for others is commonly observed in Williams syndrome patients. This rare neurodevelopmental disorder is caused by a deletion of about 26 genes from the long arm of chromosome 7 (Meyer-Lindenberg, Mervis, & Berman, 2006; Riby & Back, 2010). The extent of cognitive empathy in individuals with Williams syndrome is difficult to judge due to comorbid cognitive disability. However, they have an ability to manipulate social situations, make social references and discriminate in false-belief tasks (for review see Barak & Feng, 2016).

While the simplicity of the categorization proposed by Smith (2006) is appealing, it does not reflect the complexity of the associated conditions or the dynamic interactions between empathy-relevant neuronal circuits. The neuronal circuits involved in cognitive and emotional empathy are interconnected. They also both send and receive projections to and from many other brain regions, which can be independently affected in the conditions described above (Lamm, Decety, & Singer, 2011). Smith (2006) also does not take into account the full spectrum of neurological cases with comorbid empathy impairments. Focal lesions in the medial prefrontal cortex and inferior frontal gyrus as well as temporal pole and anterior insula can reproduce a lack of either cognitive or emotional empathy (Leigh et al., 2013; Shamay-Tsoory, 2011). Similarly, fronto-temporal lobar degeneration and Parkinson's diseases were found to impair both cognitive and emotional aspects of empathy (Cerami et al., 2014; Narme et al., 2013). Facial emotion recognition was also recently found disrupted in individuals with agenesis of the corpus callosum (Bridgman et al., 2014) and a behavioral variant of fronto-temporal dementia (characterized by a diminished response to the needs and feelings of others and/or diminished social interest as well as perseverative, stereotyped behavior (Bora, Velakoulis, & Walterfang, 2016), while perspective taking was diminished in subjects with depression (Domes et al., 2016). Selective impairment of affective empathy has also been observed in cocaine addicts (Preller et al., 2014). Idiopathic generalized epilepsy (Jiang et al., 2014) and stroke (Yeh and Tsai, 2014) were also associated with decreased empathy, with the extent of the disruption depending on the brain area and size of the affected region.

Moreover, the disorders referred to by Smith (2006) often represent a wide spectrum of cases (e.g., ASD). Subjects characterized with impairment of only one type of empathy (cognitive in the case of ASD) fall within that spectrum. The more severe cases, however, are likely to display more generalized empathic deficit. Also comorbid conditions (e.g., epilepsy) need to be taken into account due to their effect on empathic responsiveness.

With such heterogeneous etiology, finding a common molecular background for empathy disorders will be very difficult at best.

MOUSE MODELS OF EMPATHY DEFICITS

Modeling of empathy deficits in neuropsychiatric disorders, done for the purpose of detailed analysis of neuronal circuitry involved, requires the use of either genetic models of monogenic forms of these disorders or well-validated models of an idiopathic form of the disorder. Several such models are available for ASD (for a review see Ebrahimi-Fakhari & Sahin, 2015) and Williams syndrome (Osborne, 2010). Much less, however, is known about the genetic causes of schizophrenia (Escudero & Johnstone, 2014) or psychopathy (Cumming, 2015). Until more is known about these conditions, the use of animal models of these disorders is less instructive.

The use of rodents to study empathy is still the subject of scientific dispute, partly due to the lack of agreement about attribution of complex cognitive empathic abilities to animals other than humans. In his influential article, de Waal (2008) proposed a multilevel model of empathy, which set emotional contagion (observed in most, if not all, mammalian species) as the simplest and most primary form of empathy, followed by sympathetic concern (displayed by monkeys and humanoid apes) and finally by perspective taking abilities. He suggested that perspective taking might require the presence of a particular type of neurons (von Economo cells) in the anterior cingulate and anterior insular cortices. These cells have been identified in humans, apes, certain cetaceans, and African as well as Asian elephants (Allman et al., 2010). The numbers of von Economo cells were altered in many disorders characterized with empathy deficiency. In ASD their numbers were increased (Santos et al., 2011), while in early onset schizophrenia a lower density of von Economo cells correlated with the duration of sickness (Brüne et al., 2010). Reduced numbers of these cells were also reported for subjects with Alzheimer's disease (Nimchinsky, Vogt, Morrison, & Hof, 1995) and fronto-temporal dementia (Seeley et al., 2006).

The ability to empathize was also linked with the activity of the Mirror Neuron System (for details see Chapters 4 and 6). Unfortunately, no evidence for the existence of either the von Economo cells or a functional Mirror Neuron System is available in rodents. It is therefore possible that other neuronal circuits are responsible for emotional contagion and other empathic responses in these animals.

An alternative model of empathy proposed by Panksepp and Panksepp (2013) suggests that deep subcortical structures (responsible for emotional contagion and primary emotions) as well as basal ganglia and limbic structures (involved in learning and memory of shared emotions) may form the "empathy-circuitry" in animals other than humans. Refined cognitive empathy, observed in humans and their closest relatives, would require top-down control from cortical, empathy-relevant regions.

Differences in the size and activity of the amygdalar complex were found in individuals with empathy deficits. It is reduced in individuals with psychopathic trait (Marsh, 2015) and its connectivity and cell composition is altered in ASD (Morgan, Barger, Amaral, & Schumann, 2014). In line with the increased empathic affect, the size of the amygdala is enhanced in Williams syndrome patients (Capitão et al., 2011).

In rodents, amygdalar function can be altered either with the use of directed, conditioned genetic modifications or by transient inactivation or over stimulation of this brain region. Validated empathy-relevant behavioral paradigms are, however, required to assess the effects of such modifications on empathic behaviors. In recent years several protocols of this sort have been developed, most of which are described in detail in other chapters of this book. Here, I will discuss only the observations relevant to disorders characterized by impaired empathic responses. As this is a novel avenue of empathy research, there are very few studies directly addressing this issue with the use of mouse models.

Observational learning was demonstrated in highly social C57BL6/J (B6) mice (Jeon et al., 2010) in a paradigm requiring one mouse (the Observer) to witness fear conditioning training of another mouse (the Demonstrator). The formation of the freezing response in the Observer depended on the activity of $Ca(v)1.2$ Ca^{2+}channels within the anterior cingulate cortex (Jeon et al., 2010), and their local deletion in that region resulted in a lack of observational fear learning. Similarly, local inhibition of D_2 dopamine receptors, as well as administration of serotonin or a mix of dopamine and serotonin, into the anterior cingulate cortex reduced observational freezing (Kim et al., 2014). Inactivation of the anterior cingulate cortex and parafascicular or mediodorsal thalamic nuclei (responsible for pain affection) also resulted in decreased freezing in the Observer mice (Jeon et al., 2010). The activity of neurons in the anterior cingulate cortex of these mice was synchronized (in the theta band frequency) with the activity of neurons in the lateral nucleus of the amygdala during observation of the conditioning session (Jeon et al., 2010). While administration of serotonin changed anterior cingulate cortex activity in the delta band (increase) as well as in alpha and gamma bands (decrease), infusion of dopamine reduced only alpha band activity (Kim et al., 2014).

Acquisition of observational fear depends on mouse strain. Chen, Panksepp, and Lahvis (2009) showed that the response to conditioned stimuli associated with the observed fear response of another mouse, namely freezing and changes in heart rate, was greater in the highly social B6 strain, than in the less gregarious BALB/c mice. They also showed that exposure to ultrasonic vocalizations recorded during the fear conditioning session was sufficient to elicit a freezing response in the B6, but not in the BALB/c strain. A recent study by Keum and coworkers (2016) expanded on that topic and showed that while C57BL/6J, C57BL/6NTac, 129S1/SvImJ, 129S4/SvJae, and BTBR T + Itpr3tf/J(BTBR) mice do show observational fear learning, AKR/J, BALB/cByJ, C3H/HeJ, DBA/2J, FVB/NJ, and NOD/ShiLtJ do not display that response.

Contrary to this last finding, BTBR mice (a well-validated mouse model of idiopathic autism, for review see Meyza and Blanchard, 2017) were recently found to lack the empathic between-subject transfer of emotional information (Meyza et al., 2015). They also did not display a characteristic increase in neuronal activation in the amygdala and prefrontal cortex resulting from that transfer.

The difference in the outcome of these two studies likely stems from the use of distinct experimental protocols. While observational fear learning requires the Observer mouse to directly witness the fear conditioning training of a familiar individual (unfamiliar mice do not elicit such reaction, Gonzalez-Liencres, Juckel, Tas, Friebe, & Brüne, 2014; Jeon et al., 2010), between-subject transfer of emotional information relies on information exchange upon reunion of the cagemates (the Demonstrator and the Observer) following the exposure of the Demonstrator to a fear conditioning session (Knapska et al., 2006). In a way, the latter protocol is closer to the definition of empathy as a prosocial drive (Decety et al., 2016). It focuses not just on the buildup of arousal in the Observer (an equivalent of personal distress reported by human Observers during pain-viewing tasks), but allows for the occurrence of spontaneous consolatory behavior in the safe environment of the home cage. Moreover, observational fear learning requires previous conditioning experience of the Observers (Sanders, Mayford, & Jeste, 2013), while no prior training is required for the between-subject transfer of emotional information test. Observation of pro-social behaviors in the latter protocol can therefore be attributed to the emotional contagion alone, while freezing in the observational fear learning protocol may result, at least in part, from context to cue-based fear renewal.

The difference in response to imminent versus remote information about danger was also demonstrated for autism-relevant mouse strains using shared experience paradigms (where two animals are subjected to conditioning at the same time). Co-learning of contextual fear responses was enhanced by the presence of a BTBR individual in the conditioning chamber both in B6 and BTBR mice (Lipina & Roder, 2013). Similarly, enhanced pain response was observed in BALB/c mice when paired with another BALB/c mouse in pain (Langford et al., 2006). On the other hand, co-learning of cue-based fear was enhanced only in the B6 strain and not in the BTBR mice (Lipina & Roder, 2013).

In sum, deficient empathic responsiveness was demonstrated, using several protocols, in select mouse models characterized by low sociability. Further studies are required to verify these findings in mouse models of monogenic forms of ASD, as well as in mouse models of other neurological and neuropsychiatric disorders with comorbid empathy deficits.

OXYTOCIN AS A THERAPEUTIC AGENT

The use of rodent models of disorders characterized with deficient empathy offers the possibility of preclinical testing of novel therapeutic agents as well as an insight into the molecular background and long-term effects of already existing therapeutic strategies.

One such treatment involves the use of exogenous oxytocin to modulate social behavior. Polymorphisms in oxytocin-receptor allele were linked to ASDs and associated with a decreased size of the anterior cingulate cortex (Tost et al., 2010). Oxytocin receptors are located in many empathy-relevant brain regions, including the prelimbic cortex, the nucleus accumbens, and the amygdala (human: Boccia, Petrusz, Suzuki, Marson, & Pedersen, 2013; mouse: Gould & Zingg, 2003). A recent study showed that altering oxytocin neurotransmission in the mother's central amygdala affects the rat mother-to-pup transfer of emotional information (Rickenbacher et al. 2017). Intranasal oxytocin administration was associated with reduced blood oxygen level-dependent (BOLD) signals in the dorsal striatum and the amygdalar complex (Baumgartner, Heinrichs, Vonlanthen, Fischbacher, & Fehr, 2008; Kirsch et al., 2005).

While some studies showed enhanced trust and cooperation as well as improved social skills in individuals with autism exposed to intranasal oxytocin, others found weak or no effects for that treatment (for review see Young and Barrett 2015). In borderline patients, oxytocin supplementation actually worsened social anxiety (Bartz et al., 2011).

The results of oxytocin administration in animals other than humans are also equivocal. In macaques, intranasal supplementation increased prosocial behaviors towards another monkey only 2 h following administration. Until then it promoted self-centered actions (Chang, Barter, Ebitz, Watson, & Platt, 2012). In mice, the effect of oxytocin administration depended on the type of exposure (intranasal vs. intraperitoneal, *i.p.*) and the timing regimen (acute, subchronic, or chronic) used. Teng and collaborators (2013) showed that while acute *i.p.* administration of oxytocin did not improve sociability in BALB/cByJ mice, subchronic (4 times over 9 days) supplementation did so already 24 h after cessation of treatment. The same effect in the C58/Jmales had a delayed onset and was present only 2 weeks after the end of the treatment. At the same time, stereotypy was reduced, and grooming enhanced in C58/J mice, by acute rather than subchronic administration (Teng et al., 2013). Subchronic *i.p.* administration was also efficient at reducing social deficits in Grin1 (Glutamate Ionotropic Receptor N-methyl-D-aspartate (NMDA) Type Subunit 1) knockdown mice (Teng et al., 2016). The effects of intranasal supplementation of oxytocin, much like the *i.p.* administration, depended on the duration of the treatment. Unlike the *i.p.* administration, acute intranasal oxytocin enhanced social contacts, but only with mice of the opposite sex. Chronic treatment, on the other hand, impaired sociability of B6 mice

(Huang et al., 2014), failed to improve sociability of the BTBR mice (Bales et al., 2014) and resulted in deficient pair bonding in prairie voles (Bales et al., 2013). In the ketamine-induced model of schizophrenia, subchronic (4 times over 9 days) oxytocin failed to improve sociability in the ketamine treated animals, but did so in ketamine free animals (Sobota, Mihara, Forrest, Featherstone, & Siegel, 2015).

In sum, despite the obvious link between oxytocin signaling and the sociability of both humans and nonhuman animals, the use of exogenous oxytocin to alleviate social deficits in neuropsychiatric and neurological disorders characterized with impaired empathy requires further testing. The route of delivery, administration regime, and individual responsiveness to a treatment all need to be taken into account prior to commencement of oxytocin supplementation.

SUMMARY

The development of behavioral paradigms for studying empathy deficits in rodents opens an entirely new avenue of research. The choice of methods, however, should be made carefully and with the specific type of empathy impairment to be studied in mind. It has become clear that direct observation or sharing of an aversive experience can elicit emotional contagion and enhanced fearful response in mice. However, less challenging social encounters with a distressed conspecific in a neutral environment may elicit more subtle and nuanced empathic and prosocial responses. Further studies are needed to confirm if other mouse models of human neuropsychiatric disorders reproduce the lack of empathic responses displayed by relevant human subjects. If so, these models will be of great value for deciphering the genetic and epigenetic causes of empathy impairments in many neurological disorders.

References

Allman, J. M., Tetreault, N. A., Hakeem, A. Y., Manaye, K. F., Semendeferi, K., Erwin, J. M., et al. (2010). The von economo neurons in frontoinsular and anterior cingulate cortex in great apes and humans. *Brain Structure and Function, 214*(5–6), 495–517.

Bales, K. L., Perkeybile, A. M., Conley, O. G., Lee, M. H., Guoynes, C. D., Downing, G. M., et al. (2013). Chronic intranasal oxytocin causes long-term impairments in partner preference formation in male prairie voles. *Biological Psychiatry, 74*(3), 180–188.

Bales, K. L., Solomon, M., Jacob, S., Crawley, J. N., Silverman, J. L., Larke, R. H., et al. (2014). Long-term exposure to intranasal oxytocin in a mouse autism model. *Translational Psychiatry, 4*, e480.

Barak, B., & Feng, G. (2016). Neurobiology of social behavior abnormalities in autism and Williams syndrome. *Nature Neuroscience, 19*(6), 647–655.

Bartz, J., Simeon, D., Hamilton, H., Kim, S., Crystal, S., Braun, A., et al. (2011). Oxytocin can hinder trust and cooperation in borderline personality disorder. *Social Cognitive and Affective Neuroscience, 6*(5), 556–563.

Baumgartner, T., Heinrichs, M., Vonlanthen, A., Fischbacher, U., & Fehr, E. (2008). Oxytocin shapes the neural circuitry of trust and trust adaptation in humans. *Neuron, 58,* 639–650.

Bernhardt, S. L., Valk, G., Silani, G., Bird, U., Frith, U., & Singer, T. (2014). Selective disruption of sociocognitive structural brain networks in autism and alexithymia. *Cerebral Cortex, 24*(12), 3258–3267.

Berthoz, S., & Hill, E. L. (2005). The validity of using self-reports to assess emotionregulation abilities in adults with autism spectrum disorder. *EuropeanPsychiatry, 20*(3), 291–298.

Bird, G., Silani, G., Brindley, R., White, S., Frith, U., & Singer, T. (2010). Empathic brain responses in insula are modulated by levels of alexithymia but not autism. *Brain, 133*(Pt 5), 1515–1525.

Blair, R. J. R. (2013). The neurobiology of psychopathic traits in youths. *Nature Reviews Neuroscience, 14*(11), 786–799.

Boccia, M. L., Petrusz, P., Suzuki, K., Marson, L., & Pedersen, C. A. (2013). Immunohistochemical localization of oxytocin receptors in human brain. *Neuroscience, 253,* 155–164.

Bora, E., Velakoulis, D., & Walterfang, M. (2016). Meta-analysis of facial emotion recognition in behavioral variant frontotemporal dementia: Comparison with Alzheimer disease and healthy controls. *Journal of Geriatric Psychiatry and Neurology, 29*(4), 205–211.

Brewer, R., Biotti, F., Catmur, C., Press, C., Happé, F., Cook, R., et al. (2016). Can neuro typical individuals read autistic facial expressions? A typical production of emotional facial expressions in autism spectrum disorders. *Autism Research, 9*(2), 262–271.

Bridgman, M. W., Brown, W. S., Spezio, M. L., Leonard, M. K., Adolphs, R., & Paul, L. K. (2014). Facial emotion recognition in agenesis of the corpus callosum. *Journal of Neurodevelopment Disorders, 6*(1), 32.

Brüne, M., Schöbel, A., Karau, R., Benali, A., Faustmann, P. M., Juckel, G., et al. (2010). Von Economo neuron density in the anterior cingulate cortex is reduced in early onset schizophrenia. *Acta Neuropathologica, 119*(6), 771–778.

Capitão, L., Sampaio, A., Sampaio, C., Vasconcelos, C., Férnandez, M., Garayzábal, E., et al. (2011). MRI amygdala volume in Williams syndrome. *Research in Developmental Disabilities, 32*(6), 2767–2772.

Cerami, C., Dodich, A., Canessa, N., Crespi, C., Marcone, A., Cortese, F., et al. (2014). Neural correlates of empathic impairment in the behavioral variant of frontotemporal dementia. *Alzheimer's & Dementia, 10*(6), 827–834.

Chang, S. W., Barter, J. W., Ebitz, R. B., Watson, K. K., & Platt, M. L. (2012). Inhaled oxytocin amplifies both vicarious reinforcement and self reinforcement in rhesus macaques (Macaca mulatta). *Proceedings of the National Academy of Sciences of the United States of America, 109*(3), 959–964.

Chen, Q., Panksepp, J. B., & Lahvis, G. P. (2009). Empathy is moderated by genetic background in mice. *PLoS One, 4*(2), e4387.

Cook, R., Brewer, R., Shah, P., & Bird, G. (2013). Alexithymia, not autism, predicts poor recognition of emotional facial expressions. *Psychological Science, 24*(5), 723–732.

Crawford, H., Moss, J., Anderson, G. M., Oliver, C., & McCleery, J. P. (2015). Implicit discrimination of basic facial expressions of positive/negative emotion in fragile x syndrome and autism spectrum disorder. *American Journal on Intellectual and Developmental Disabilities, 120*(4), 328–345.

Cumming, M. A. (2015). The neurobiology of psychopathy: Recent developments and new directions in research and treatment. *CNS Spectrums, 20*(3), 200–206.

de Waal, F. B. (2008). Putting the altruism back into altruism: The evolution of empathy. *Annual Review of Psychology, 59,* 279–300.

Decety, J., Skelly, L. R., & Kiehl, K. A. (2013). Brain response to empathy-eliciting scenarios involving pain in incarcerated individuals with psychopathy. *JAMA Psychiatry, 70,* 638–645.

Decety, J., Bartal, I. B., Uzefovsky, F., & Knafo-Noam, A. (2016). Empathy as a driver of prosocial behaviour: Highly conserved neurobehavioural mechanisms across species. *Philosophical Transactions of the Royal Society of London B: Biological Sciences, 371*(1686), 20150077.

Derntl, B., Michel, T. M., Prempeh, P., Backes, V., Finkelmeyer, A., & Schneider, F. (2015). Empathy in individuals clinically at risk for psychosis: Brain and behaviour. *British Journal of Psychiatry, 207*(5), 407–413.

Domes, G., Spenthof, I., Radtke, M., Isaksson, A., Normann, C., & Heinrichs, M. (2016). Autistic traits and empathy in chronic vs. episodic depression. *Journal of Affective Disorders, 195*, 144–147.

Dziobek, I., Rogers, K., Fleck, S., Bahnemann, M., Heekeren, H. R., & Wolf, O. T. (2008). Dissociation of cognitive and emotional empathy in adults with Asperger syndrome using the multifaceted empathy test (MET). *Journal of Autism and Developmental Disorders, 38*(3), 464–473.

Ebrahimi-Fakhari, D., & Sahin, M. (2015). Autism and the synapse: Emerging mechanisms and mechanism-based therapies. *Current Opinion in Neurology, 28*(2), 91–102.

Escudero, I., & Johnstone, M. (2014). Genetics of schizophrenia. *Current Psychiatry Reports, 16*(11), 502.

Gonzalez-Liencres, C., Juckel, G., Tas, C., Friebe, A., & Brüne, M. (2014). Emotional contagion in mice: The role of familiarity. *Behavioural Brain Research, 263*, 16–21.

Gould, B. R., & Zingg, H. H. (2003). Mapping oxytocin receptor gene expression in the mouse brain and mammary gland using an oxytocin receptor-LacZ reporter mouse. *Neuroscience, 122*(1), 155–167.

Green, M. F., Horan, W. P., & Lee, J. (2015). Social cognition in schizophrenia. *Nature Reviews Neuroscience, 16*(10), 620–631.

Hadjikhani, N., Zürcher, N., Rogier, O., Hippolyte, L., Lemonnier, E., & Ruest, T. (2014). Emotional contagion for pain is intact in autism spectrum disorders. *Translational Psychiatry, 4*, e343.

Horan, W. P., Jimenez, A. M., Lee, J., Wynn, J. K., Eisenberger, N. I., & Green, M. F. (2016). Pain empathy in schizophrenia: An fMRI study. *Social Cognitive and Affective Neuroscience, 11*(5), 783–792.

Huang, H., Michetti, C., Busnelli, M., Managò, F., Sannino, S., Scheggia, D., et al. (2014). Chronic and acute intranasal oxytocin produce divergent social effects in mice. *Neuropsychopharmacology, 39*(5), 1102–1114.

Jeon, D., Kim, S., Chetana, M., Jo, D., Ruley, H. E., Lin, S. Y., et al. (2010). Observational fear learning involves affective pain system and Cav1 2 Ca^{2+} channels in ACC. *Nature Neuroscience, 13*(4), 482–488.

Jiang, Y., Hu, Y., Wang, Y., Zhou, N., Zhu, L., & Wang, K. (2014). Empathy and emotion recognition in patients with idiopathic generalized epilepsy. *Epilepsy & Behavior, 37*, 139–144.

Keum, S., Park, J., Kim, A., Park, J., Kim, K. K., Jeong, J., et al. (2016). Variability in empathic fear response among 11 inbred strains of mice. *Genes, Brain and Behavior, 15*(2), 231–242.

Kim, B. S., Lee, J., Bang, M., Seo, B. A., Khalid, A., Jung, M. W., et al. (2014). Differential regulation of observational fear and neural oscillations by serotonin and dopamine in the mouse anterior cingulate cortex. *Psychopharmacology (Berl), 231*(22), 4371–4381.

Kirsch, P., Esslinger, C., Chen, Q., Mier, D., Lis, S., Siddhanti, S., et al. (2005). Oxytocin modulates neural circuitry for social cognition and fear in humans. *The Journal of Neuroscience, 25*, 11489–11493.

Knapska, E., Nikolaev, E., Boguszewski, P., Walasek, G., Blaszczyk, J., Kaczmarek, L., et al. (2006). Between-subject transfer of emotional information evokes specific pattern of amygdala activation. *Proceedings of the National Academy of Sciences of the United States of America, 103*(10), 3858–3862.

Komeda, H., Kosaka, H., Saito, D. N., Mano, Y., Jung, M., Fujii, T., et al. (2015). Autistic empathy toward autistic others. *Social Cognitive and Affective Neuroscience, 10*(2), 145–152.

Krach, S., Kamp-Becker, I., Einhäuser, W., Sommer, J., Frässle, S., Jansen, A., et al. (2015). Evidence from pupillometry and fMRI indicates reduced neural response during vicarious social pain but not physical pain in autism. *Human Brain Mapping, 36*(11), 4730–4744.

Lamm, C., Decety, J., & Singer, T. (2011). Meta-analytic evidence for common and distinct neural networks associated with directly experienced pain and empathy for pain. *Neuro-Image, 54*, 2492–2502.

Langford, D. J., Crager, S. E., Shehzad, Z., Smith, S. B., Sotocinal, S. G., Levenstadt, J. S., et al. (2006). Social modulation of pain as evidence for empathy in mice. *Science, 312*(5782), 1967–1970.

Leigh, R., Oishi, K., Hsu, J., Lindquist, M., Gottesman, R. F., Jarso, S., et al. (2013). Acute lesions that impair affective empathy. *Brain, 136*(Pt 8), 2539–2549.

Lipina, T. V., & Roder, J. C. (2013). Co-learning facilitates memory in mice: A new avenue in social neuroscience. *Neuropharmacology, 64*, 283–293.

Lockwood, P. L. (2016). The anatomy of empathy: Vicarious experience and disorders of social cognition. *Behavioural Brain Research, 311*, 255–266.

Lockwood, P. L., Sebastian, C. L., McCrory, E. J., Hyde, Z. H., Gu, X., De Brito, S. A., et al. (2013). Association of callous traits with reduced neural response to others' pain in children with conduct problems. *Current Biology, 23*, 901–905.

Lombardo, M. V., Chakrabarti, B., Bullmore, E. T., et al. (2010). Atypical neural self-representation in autism. *Brain, 133*, 611–624.

Marsh, A. A. (2015). Understanding amygdala responsiveness to fearful expressions through the lens of psychopathy and altruism. *Journal of Neuroscience Research*. doi: 10.1002/jnr.23668.

Marsh, A. A., Finger, E. C., Fowler, K. A., Adalio, C. J., Jurkowitz, I. T., Schechter, J. C., et al. (2013). Empathic responsiveness in amygdala and anterior cingulate cortex in youths with psychopathic traits. *Journal of Child Psychology and Psychiatry, 54*(8), 900.

Meffert, H., Gazzola, V., den Boer, J. A., Bartels, A. A., & Keysers, C. (2013). Reduced spontaneous but relatively normal deliberate vicarious representations in psychopathy. *Brain, 136*(Pt 8), 2550–2562.

Meyer-Lindenberg, A., Mervis, C. B., & Berman, K. F. (2006). Neural mechanisms in Williams syndrome: A unique window to genetic influences on cognition and behaviour. *Nature Reviews Neuroscience, 7*(5), 380–393.

Meyza, K. Z., & Blanchard, D. C. (2017). The BTBR mouse model of idiopathic autism: current view on mechanisms. *Neuroscience Biobehavioral Review, 276*(Pt A), 99–110.

Meyza, K., Nikolaev, T., Kondrakiewicz, K., Blanchard, D. C., Blanchard, R. J., & Knapska, E. (2015). Neuronal correlates of asocial behavior in a BTBR T (+) Itpr3(tf)/J mouse model of autism. *Frontiers in Behavioral Neuroscience, 9*, 199.

Michalska, K. J., Zeffiro, T. A., & Decety, J. (2016). Brain response to viewing others being harmed in children with conduct disorder symptoms. *Journal of Child Psychology and Psychiatry, 57*(4), 510–519.

Minio-Paluello, I., Baron-Cohen, S., Avenanti, A., Walsh, V., & Aglioti, S. M. (2009). Absence of embodied empathy during pain observation in Asperger syndrome. *Biological Psychiatry, 65*(1), 55–62.

Minio-Paluello, I., Lombardo, M. V., Chakrabarti, B., Wheelwright, S., & Baron-Cohen, S. (2009). Response to smith's letter to the editor emotional empathy in autism spectrum conditions: Weak, intact, or heightened? *Journal of Autism and Developmental Disorders*. doi: 10.1007/s10803-009-0800-x.

Morgan, J. T., Barger, N., Amaral, D. G., & Schumann, C. M. (2014). Stereological study of amygdala glial populations in adolescents and adults with autism spectrum disorder. *PLoS One, 9*(10), e110356.

Mothersill, O., Knee-Zaska, C., & Donohoe, G. (2016). Emotion and theory of mind in schizophrenia-investigating the role of the cerebellum. *Cerebellum, 15*(3), 357–368.

Narme, P., Mouras, H., Roussel, M., Duru, C., Krystkowiak, P., & Godefroy, O. (2013). Emotional and cognitive social processes are impaired in Parkinson's disease and are related to behavioral disorders. *Neuropsychology, 27*(2), 182–192.

Nimchinsky, E. A., Vogt, B. A., Morrison, J. H., & Hof, P. R. (1995). Spindle neurons of the human anterior cingulate cortex. *The Journal of Comparative Neurology, 355*(1), 27–37.

Osborne, L. R. (2010). Animal models of Williams syndrome. *American Journal of Medical Genetics part C Seminars in Medical Genetics, 154C*(2), 209–219.

Panksepp, J., & Panksepp, J. B. (2013). Toward a cross-species understanding of empathy. *Trends in Neurosciences, 36*(8), 489–496.

Pfeifer, J. H., Merchant, J. S., Colich, N. L., Hernandez, L. M., Rudie, J. D., & Dapretto, M. (2013). Neural and behavioral responses during self-evaluative processes differ in youth with and without autism. *Journal of Autism and Developmental Disorders, 43*, 272–285.

Preller, K. H., Hulka, L. M., Vonmoos, M., Jenni, D., Baumgartner, M. R., Seifritz, E., et al. (2014). Impaired emotional empathy and related social network deficits in cocaine users. *Addiction Biology, 19*(3), 452–466.

Press, C., Richardson, D., & Bird, G. (2010). Intact imitation of emotional facial actions in autism spectrum conditions. *Neuropsychologia, 48*(11), 3291–3297.

Riby, D. M., & Back, E. (2010). Can individuals with Williams syndrome interpret mental states from moving faces? *Neuropsychologia, 48*(7), 1914–1922.

Rickenbacher, E., Perry, R. E., Sullivan, R. M., & Moita, M. A. (2017). Freezing suppression byoxytocin in central amygdala allows alternate defensive behaviours and mother-pupinteractions. *Elife, 13*(6), e24080.

Sanders, J., Mayford, M., & Jeste, D. (2013). Empathic fear responses in mice are triggered by recognition of a shared experience. *PLoS One, 8*(9), e74609.

Santos, M., Uppal, N., Butti, C., Wicinski, B., Schmeidler, J., Giannakopoulos, P., et al. (2011). Von economo neurons in autism: A stereologic study of the frontoinsular cortex in children. *Brain Research, 1380*, 206–217.

Saxe, R., & Powell, L. J. (2006). It's the thought that counts: Specific brain regions for one component of theory of mind. *Psychological Science, 17*(8), 692–699.

Seeley, W. W., Carlin, D. A., Allman, J. M., Macedo, M. N., Bush, C., Miller, B. L., et al. (2006). Early frontotemporal dementia targets neurons unique to apes and humans. *Annals of Neurology, 60*(6), 660–667.

Shamay-Tsoory, S. G. (2011). The neural bases for empathy. *Neuroscientist, 17*(1), 18–24.

Silani, G., Bird, G., Brindley, R., Singer, T., Frith, C., & Frith, U. (2008). Levels of emotional awareness and autism: An fMRI study. *Social Neuroscience, 3*(2), 97–112.

Smith, A. (2006). Cognitive empathy and emotional empathy in human behavior and evolution. *The Psychological Record, 56*, 3–21.

Smith, A. (2009). Emotional empathy in autism spectrum conditions: Weak, intact, or heightened? *Journal of Autism and Developmental Disorders, 39*(12), 1747–1748.

Sobota, R., Mihara, T., Forrest, A., Featherstone, R. E., & Siegel, S. J. (2015). Oxytocin reduces amygdala activity, increases social interactions, and reduces anxiety-like behavior irrespective of NMDAR antagonism. *Behavioral Neuroscience, 129*(4), 389–398.

Teng, B. L., Nonneman, R. J., Agster, K. L., Nikolova, V. D., Davis, T. T., Riddick, N. V., et al. (2013). Prosocial effects of oxytocin in two mouse models of autism spectrum disorders. *Neuropharmacology, 72*, 187–196.

Teng, B. L., Nikolova, V. D., Riddick, N. V., Agster, K. L., Crowley, J. J., Baker, L. K., et al. (2016). Reversal of social deficits by subchronic oxytocin in two autism mouse models. *Neuropharmacology, 105*, 61–71.

Tost, H., Kolachana, B., Hakimi, S., Lemaitre, H., Verchinski, B. A., Mattay, V. S., et al. (2010). A common allele in the oxytocin receptor gene (OXTR) impacts prosocial temperament and human hypothalamic-limbic structure and function. *Proceedings of the National Academy of Sciences of the United States of America, 107*, 13936–13941.

Usui, S., Senju, A., Kikuchi, Y., Akechi, H., Tojo, Y., Osanai, H., et al. (2013). Presence of contagious yawning in children with autism spectrum disorder. *Autism Research and Treatment, 2013*, 971686.

Walsh, J. A., Creighton, S. E., & Rutherford, M. D. (2016). Emotion perception or social cognitive complexity: What drives face processing deficits in autism spectrum disorder? *Journal of Autism and Developmental Disorders*, *46*(2), 615–623.

Yeh, Z. T., & Tsai, C. F. (2014). Impairment on theory of mind and empathy in patients with stroke. *Psychiatry and Clinical Neurosciences*, *68*(8), 612–620.

Young, L. J., & Barrett, C. E. (2015). Neuroscience. Can oxytocin treat autism? *Science*, *347*(6224), 825–826.

15

Future Directions, Outstanding Questions

Ewelina Knapska, Ksenia Z. Meyza

Nencki Institute of Experimental Biology, Polish Academy of Sciences, Warsaw, Poland

The acknowledgement of the evolutionary roots of empathy, and the existence of some of its aspects in animals other than humans, forms a framework for studying the neuronal basis of this phenomenon in unprecedented detail. When Preston and De Waal (2002) and De Waal (2008) proposed the unified theory of empathy, they opened the door to studying the neuronal mechanisms of empathic responsiveness in animal models. The recent dramatic increase in the number of studies of prosocial behavior in rodents and primates lends credence to the theory of the evolutionary roots of empathy (Chapters 5 and 13). These studies have shown that rodents are capable of emotional contagion and that the presence of a fearful partner or a familiar individual in pain results in induction of similar emotions in the witnessing conspecifics (Chapters 8–10). Further, emotional contagion can compel rodents to engage in prosocial behaviors (Chapter 12). Pair bonding (Chapter 7) and social buffering (Chapter 11) phenomena have also been studied in depth, offering the possibility of translational studies within these domains. The role of brain regions such as the amygdala and the medial prefrontal cortex in the shaping of empathic responses has also recently been confirmed for both rodents and humans. These recent theoretical and experimental advancements suggest the great potential of using animal models in research of neuronal correlates of empathy.

The need for a detailed, mechanistic insight into the neurobiology of empathy originates from the poor spatial and temporal resolution of neuroimaging techniques currently used to study the neuronal correlates of vicarious experiences in humans. These studies have provided a wealth of data on brain regions involved in shared representations of emotions

Neuronal Correlates of Empathy. http://dx.doi.org/10.1016/B978-0-12-805397-3.00015-2

(Chapters 2 and 3). However, in the light of the recent shift in recognition of functional units in the brain—from whole brain structures to specific circuits, which often colocalize with other, functionally adverse circuits within the same brain region—our understanding of the neuronal background of empathy relies on the proper recognition of the phenotype of neurons involved in the process. While single-unit electrophysiological recordings from brain regions previously identified as empathy-relevant could provide part of that information, they are primarily reserved for animal research. Moreover, unless multiple recording sites are synchronously recorded from one individual, they are unlikely to provide information about the broader circuitry. Such complex information is likely to be obtained through a combination of chemogenetic or optogenetic tools along with modern in vivo imaging techniques, such as the frame-projected independent fiber-photometry imaging based on genetically encoded calcium indicators.

Newly developed technologies for manipulation of neuronal activity provide unique opportunities to investigate previously difficult or impossible to solve problems. The list of such problems includes the most fundamental questions—Can social processes be reduced to nonsocial ones? Are there neural systems unique to the social domain or are general neural systems involved in the processing of both social and nonsocial stimuli? (cf. Adolphs, 2003). In the past years, we have learned that the anterior temporal lobes, several nuclei of the amygdala, and the medial/orbital prefrontal cortex take part in perceiving social stimuli and producing appropriate behavioral responses. These structures show consistent activation in response to both social and nonsocial tasks. As the current neuroimaging techniques do not have sufficient resolution to pinpoint activation of specific neuronal circuits within a given region, presentation of social and nonsocial stimuli may appear to activate the same brain areas, even though distinct (although virtually overlapping) circuits may be in operation (Knapska et al., 2012).

Another question crucial to understanding the neural mechanisms of empathy is whether mirror neurons control social emotions. The involvement of mirror neurons has been hypothesized for a wide range of abilities, including imitation, mimicry, and empathy (Ferrari & Rizzolatti, 2014, Chapters 4 and 6). In humans, neuroimaging studies have shown mimicry-induced activation in the premotor cortex, the supplementary motor area, the primary somatosensory cortex, and the inferior parietal cortex. These activations were attributed to the action of a mirror neuron system (see Chapter 2). However, most of the results supporting the mirror mechanism were obtained with methods that do not allow identification of single cell activity. There is no direct evidence for the existence of mirror neurons in humans or rodents. Studies employing single-cell activity recordings have shown that mirror neurons constitute less than 17% of all recorded cells (Gallese, Fadiga, Fogassi, & Rizzolatti, 1996; Mukamel, Ekstrom, Kaplan, Iacoboni, & Fried, 2010). With such a low percentage, techniques measuring activation of whole brain structures cannot hope to verify whether mirror neurons are actually involved in a given process. Another group of neurons that are thought to be involved in empathizing are von Economo (spindle) neurons. The von Economo cells are located in the anterior cingulate cortex and the fronto-insular cortex in humans, apes, macaques, elephants, certain cetaceans, and raccoons. Their function is not well understood, but due to their size and abundance in the brains of highly social species with well-developed cortices, they have been linked to the fast transfer of information necessary for advanced social skills (Butti, Sherwood, Hakeem, Allman, & Hof, 2009). To understand which types of neurons are involved, how they are interconnected, and what is their function in the control of empathic behavior, techniques allowing manipulation of activity at a single-cell resolution are needed. For example, the ability to localize and block the neuronal circuits involved in emotional contagion would also illuminate whether or not higher empathic functions (e.g., helping behavior) are also affected. If the mechanisms of emotional contagion and helping behavior are distinct, the latter would not be disturbed. Certainly, such manipulations are not available in human subjects. They are, however, routinely used in rodent models.

The use of state of the art neurobiology tools along with rodent models allows investigation of empathy at a previously unattainable level. At the same time, it raises the question of the translational value of animal research. How much similarity is there between how certain neurocognitive processes function in humans compared to animals. The theory formalized by Preston and de Waal assumes that emotional contagion is a mechanism on which more advanced empathic behaviors are built. However, there is no consensus about the role of emotional contagion in empathy (Chapters 3 and 4). To resolve this problem, we need to broaden our understanding of the evolutionary continuity of empathic responses. Studies dedicated to this topic, however, need to be done with utmost care to avoid unwarranted anthropomorphism.

Both human and animal studies of empathy need to take into account many factors affecting emotional reactivity of the subjects. These include sex/gender, stress level, the character of the vicarious experience (imminent/remote stressor exposure), the nature of the stimulus, and the familiarity bias. In particular, future studies should establish whether the empathy facilitating effect of familiarity develops in the same or in different ways for specific people, such as loved ones, or cage mates in laboratory animals, and groups of people we may or may not identify with or groups of animals that may or may not seem familiar. The direct, between-species comparison of empathic responses requires the use of uniform measures of stress. Physiological markers, such as blood or saliva corticosterone levels, elevated heart rate, and body temperature or skin conductance, could be used for comparing the dynamics of the autonomic component of empathic responses between species. Such unbiased data would help us recognize the best time windows for actions such as tissue collection for detailed analysis of neuronal activation resulting from the behavioral arousal. Many of the current misunderstandings and/or seemingly contradictory results in the field of animal empathy research arise from application of different behavioral protocols for testing supposedly uniform empathic abilities. It is also important to acknowledge that there is species specificity to behavioral responses to different stressors and the role of the modality of information received by the subject. For instance, the same stressful stimulus is likely to induce different defensive behaviors in rat and mouse Demonstrators. The Observers from these species would then receive a different type of information to begin with and would adjust their own behavior accordingly.

Apart from investigating the evolutionary continuity of empathic responses, it is crucial to establish the neuronal and genetic underpinnings of individual differences in empathic abilities. Although there is a considerable amount of data about structural correlates of such variability coming from neuroimaging studies (for review, see Allen et al., 2017), the genetic background of individual differences in empathizing is less well understood. Within the human population, genetic studies of certain subpopulations, lacking either cognitive or emotional empathy, could pinpoint the mutations responsible for decreased empathic responsiveness. To verify their role in the development of neuropsychiatric disorders characterized with deficient empathy, relevant animal models need to be used. Recent advance in genetic engineering has provided tools for relatively easy gene editing (e.g., the Clustered Regularly Interspaced Short Palindromic Repeats, commonly known as CRISP/Cas technology, Bell et al., 2016). And although this technology is currently used mostly in laboratory rodents, first studies using neurons developed from induced pluripotent stem cells obtained from patients with neurological disorders report rescue of the phenotype of these cells with gene editing treatment (e.g., Zhang et al., 2017). It is difficult to say how far we are at this point from being able to safely restore function in human neurons (or stem cells, which upon infusion back into the brain of the patient would safely differentiate into healthy neurons). It is, however, likely that with time, safe protocols for such intervention will be developed.

Another open research topic is the formation of empathic responsiveness during ontogeny. While it has been studied extensively in children (for review, see Decety, Bartal, Uzefovsky, & Knafo-Noam, 2016), little is known about the unveiling of empathy in young primates and rodents.

A recent study showed that transfer of fear to a novel odor from rat mothers to pups was dependent on chemosignaling and the activity of the basal and lateral amygdala (Debiec & Sullivan, 2014). Blocking of oxytocin signaling in the central amygdala of the mothers in that paradigm changes their defensive responses, which leads to disruption of the fear transfer (Rickenbacher et al. 2017). Further research is needed to see if other forms of empathy (especially prosocial behaviors) develop at a similar developmental stage and whether early transfer of emotions is present in other laboratory rodent species. This is of special importance for mice, as many models of genetic and idiopathic forms of neurodevelopmental disorders with comorbid empathy deficits are developed in this species. Understanding the stages by which emotional contagion develops in mice would improve the translatability of research done on mouse models of neurodevelopmental disorders, allowing comparative studies with affected children.

In sum, the development of standard protocols for assessment of different forms of empathy would greatly improve the comparability of data obtained across different species and thus our understanding of the evolutionary continuity of the phenomenon. The genetic background of individual empathic responsiveness and of empathy deficits observed in multiple neuropsychiatric disorders is not well understood and requires further studies. To facilitate the generation of new therapeutic protocols improving empathic responsiveness, genetically modified animal models carrying relevant mutations need to be tested. At the same time, we need to use unambiguous physiological measures with little species specificity to correlate the behavioral responses with the activity of specific neuronal circuits.

The current book challenges the strictly anthropocentric view, which implies that empathy is a human-only trait, by presenting and discussing experimental data showing that animals other than humans are also capable of displaying simple forms of empathy. By admitting that animals other than humans are capable of empathy, a new and fascinating chapter of research with many unanswered questions is opened. While studies of human empathy have been and still are crucial for deciphering the neuronal background of sharing emotions, a detailed, mechanistic, and genetic analysis of the neuronal circuitry involved in empathic responsiveness, requires the use of other animals. Hopefully, the first results of such animal studies on empathy will stimulate further multidisciplinary research that will help to put the original theory of de Waal into context. We hope that by going beyond semantics and presenting the many aspects of primate and rodent empathic responses, we have managed to convince the reader of the potential for the use of these animal models in empathy research.

References

Adolphs, R. (2003). Is the human amygdala specialized for processing social information? *Annals of the New York Academy of Sciences, 985*, 326–340.

Allen, M., Frank, D., Glen, J. C., Fardo, F., Callaghan, M. F., & Rees, G. (2017). Insula and somatosensory cortical myelination and iron markers underlie individual differences in empathy. *Scientific Reports, 7*, 43316.

Bell, S., Peng, H., Crapper, L., Kolobova, I., Maussion, G., Vasuta, C., et al. (2016). A rapid pipeline to model rare neurodevelopmental disorders with simultaneous CRISPR/Cas9 gene editing. *Stem Cells Translational Medicine*doi: 10.1002/sctm.16-0158.

Butti, C., Sherwood, C. C., Hakeem, A. Y., Allman, J. M., & Hof, P. R. (2009). Total number and volume of Von Economoneurons in the cerebral cortex of cetaceans. *The Journal of Comparative Neurology, 515*(2), 243–259.

De Waal, F. B. (2008). Putting the altruism back into altruism: The evolution of empathy. *Annual Review of Psychology, 59*, 279–300.

Debiec, J., & Sullivan, R. M. (2014). Intergenerational transmission of emotional trauma through amygdala-dependent mother-to-infant transfer of specific fear. *Proceedings of the National Academy of Sciences of the United States of America, 111*(33), 12222–12227.

Decety, J., Bartal, IB-A., Uzefovsky, F., & Knafo-Noam, A. (2016). Empathy as a driver of prosocial behaviour: Highly conserved neurobehavioural mechanisms across species. *Philosophical Transactions of the Royal Society B, 371* 20150077.

Ferrari, P. F., & Rizzolatti, G. (2014). Mirror neuron research: The past and the future. *Philosophical Transactions of the Royal Society B: Biological Sciences, 369*(1644), 20130169.

Gallese, V., Fadiga, L., Fogassi, L., & Rizzolatti, G. (1996). Action recognition in the premotor cortex. *Brain, 119*(Pt 2), 593–609.

Knapska, E., Macias, M., Mikosz, M., Nowak, A., Owczarek, D., Wawrzyniak, M., et al. (2012). Functional anatomy of neural circuits regulating fear and extinction. *Proceedings of the National Academy of Sciences of the United States of America, 109*(42), 17093–17098.

Mukamel, R., Ekstrom, A. D., Kaplan, J., Iacoboni, M., & Fried, I. (2010). Single-neuron responses in humans during execution and observation of actions. *Current Biology, 20*(8), 750–756.

Preston, S. D., & De Waal, F. B. (2002). Empathy: Its ultimate and proximate bases. *Behavioral and Brain Sciences, 25*(1), 1–20.

Rickenbacher, E., Perry, R. E., Sullivan, R. M., & Moita, M. A. (2017). Freezing suppression byoxytocin in central amygdala allows alternate defensive behaviours and mother-pupinteractions. *Elife, 13*(6), e24080.

Zhang, Y., Schmid, B., Nikolaisen, N. K., Rasmussen, M. A., Aldana, B. I., Agger, M., et al. (2017). Patient iPSC-derived neurons for disease modeling of frontotemporal dementia with mutation in CHMP2B. *Stem Cell Reports* pii: S2213-6711(17)30027-9.

Index

M

Main olfactory bulb (MOB), 141
Main olfactory epithelium (MOE), 141
"Mate buffering", 138
"Maternal buffering", 138
Mating-induced aggression, 83
Mentalizing, 39
Mesolimbic dopamine circuitry, 84
Midazolam-treated rats, 157
Mimicry, 72
Mirror neurons, 3, 27–29, 39–42, 67–74, 2,
     27, 67, 69. *See also* Empathy
  advantages of, 73
  classical, 71
  role of, 27
  system, 13
Mirror self-recognition (MSR)
     test, 162
MOB. *See* Main olfactory bulb (MOB)
MOE. *See* Main olfactory epithelium
     (MOE)
Morris water maze, 154
Motor empathy, 39
Mouth mirror neurons, 71
MSR test. *See* Mirror self-recognition
     (MSR) test
Multimodal match, 70
Murhythm, 72

O

Oxytocin
  as potential moderator of stress
     effects on, 31
  as therapeutic agent, 184

P

Pain empathy
  social neuroscience of pain
     in rodents, 124
  translational neuroscience of, 127
Pair bond formation, 81, 85
PAM. *See* Perception Action Mechanism
     (PAM)
Parent-infant attachments, 79
Partner loss, 87
Partner preference test, 81
Pavlovian conditioning, 15
Perception Action Mechanism
     (PAM), 54
Prairie vole pair bond model, 80
Prediction-error learning, 14
Primary somatosensory cortex, 41

R

Rapid facial mimicry (RFM), 57
RCC. *See* rostral cingulate cortex (rCC)
Remote fear learning, in rats, neuronal
     correlates of
  emotional contagion, 111
  fear memories, social modulation
     of, 114
  socially transferred fear
     neuronal correlates of, 115
     rodent models of, 112
     sex differences in, 118
RFM. *See* Rapid facial mimicry (RFM)
Rizzolatti, Giacomo, 2
rostral cingulate cortex (rCC), 42

S

Salience network, 28
Secondary somatosensory cortex, 41
Selective aggression, 83
Semantics, empathy beyond, 1
Sexually naive prairie voles, 80, 83
Sharing emotion, 71
Social behavior
  sex differences in the neurobiology of, 81
Social buffering
  of conditioned fear responses
     neural mechanisms underlying, 141
     in rats, 140
  conditioned hyperthermia
     neural mechanisms underlying, 142
     in rats, 142
  exposure-type, examples of, 138
  housing-type, examples of, 139
  psychological factor inducing, 145
Social cognitive neuroscience, 2
Social fear-learning paradigms, 95
Socially transferred fear
  neuronal correlates of, 115
  rodent models of, 112
  sex differences in, 118
Social transmission, of associative fear
     in rodents
  direct and indirect associative fear
     learning, 93
  fear conditioning by proxy, 96
     social learning paradigm, 95
  social fear learning in rodents, 104
Somatosensory, 39
Stress
  neurobiological influence of, vole
     pair bond